Complex Numbers Made Simple

Made Simple Books

Accounting
Advertising
Auditing
Book-keeping
Business and the European
    Community
Business Communication
Business Environment, The
Business Law
Business Planning
Business Studies
Economics
English for Business
Financial Management
Human Resource Management
Information Technology
Keyboarding and Document
    Presentation
Law
Management Theory and Practice
Marketing
Office Procedures
Organizations and Management
Philosophy
Psychiatry
Psychology
Sociology
Social Services
Spreadsheet Skills (Excel)
Spreadsheet Skills (Lotus)
Statistics for Business
Teeline Shorthand
Wordprocessing Skills (WordPerfect)

Mathematics

Complex Numbers
Differentiation
Integration

Languages

French
Follow-up French (and cassette)
German
Italian
Spanish
Business French (book and cassettes)
Business German (book and cassettes)
Business Italian (book and cassettes)
Business Spanish (book only)

Computer books

Access for Windows
AmiPro for Windows
Excel for Windows
Hard Drives
The Internet
Lotus 1-2-3 (DOS)
Lotus 1-2-3 (5.0) for Windows
MS-DOS
MS-Works for Windows
Multimedia
Pageplus
Pageplus for Windows
PowerPoint
Quicken for Windows
Windows 3.1
Windows 95
Word for Windows
WordPerfect (DOS)
WordPerfect for Windows

Complex Numbers Made Simple

Verity Carr

**MADE SIMPLE
BOOKS**

Made Simple
An imprint of Butterworth-Heinemann
Linacre House, Jordan Hill, Oxford OX2 8DP
A division of the Reed Elsevier plc group

$\mathcal{R}$ A member of the Reed Elsevier plc group

OXFORD   BOSTON   JOHANNESBURG
MELBOURNE   NEW DELHI   SINGAPORE

First published 1996
Transferred to digital printing 2004
© Verity Carr 1996

British Library Cataloguing in Publication Data
A catalogue record for this book is available from the British Library.

ISBN 13: 978 0 7506 2559 3

ISBN: 0 7506 2559 7

Library of Congress Cataloguing in Publication Data
A catalogue record for this book is available from the Library of Congress.

Typeset by Laserwords

Contents

Foreword

Verity Carr is a highly talented teacher of mathematics and, although she readily admits that her real love is for the teaching of pure mathematics, she has accumulated nearly 30 years of experience teaching mathematics at all levels. Verity has a rare gift for making mathematics simple and enjoyable.

Verity holds an Honours Degree in Mathematics from the University of Wales. Her teaching experience covers the Secondary, Further and Higher Education sectors and she has been involved with the design, preparation and delivery of courses at GCSE, A level, BTEC National, and Foundation University levels. She has also spent some time in the USA teaching at the Ransom Everglades School in Miami.

At Brooklands College, Verity has taken a leading role in the development of a highly successful and well-used mathematics workshop. This facility supports students on a diverse range of courses at all levels within the college. During this period, Verity has acted as the focal point for students needing additional support in their study of mathematics. This has also allowed her to obtain first-hand experience of dealing with students' problems on a one-to-one basis.

I was delighted when Verity suggested that she could produce a comprehensive set of teaching books in mathematics. Each of these books aims to address a specific topic area in which students are presented with a learning text supported by plenty of worked examples and solutions to past examination papers.

Many students at Brooklands have derived immense benefit from Verity's work and I wholeheartedly commend this series to you. Good luck with your studies!

Mike Tooley
Dean of Technology, Brooklands College

Author's note

If you are meeting complex numbers for the first time, obviously begin at the first page and work *slowly* through the book. Do *each* exercise completely *before* looking at the worked answers. The first nine chapters cover the basics.

If you are using the book for revision, begin at the first page and travel straight through the book. You will find many useful facts on every page, and will be reminded of things you may have forgotten.

Although permission to publish this book has been obtained from the University of London Examinations and Assessment Council, it accepts no responsibility whatsoever for the accuracy or method of working in the answers given.

Although permission to publish this book has been obtained from the Associated Examining Board (AEB), the worked answers are the sole responsibility of the author.

What you must already know before starting this book

1. Quadratic equations and expressions.
2. Basic vectors.
3. Trigonometry, especially radians (radians are denoted by the symbol ^{c}).
4. Algebra–always algebra!

1 The quadratic formula

The quadratic formula for solving the quadratic equation

$$ax^2 + bx + c = 0$$

has already been studied and the solution,

$$x = \frac{-b \pm \sqrt{(b^2 - 4ac)}}{2a}$$

is already well known.

$(b^2 - 4ac)$ is called the **discriminant** of the equation.

Three possibilities can arise from the above formula:

   (i) $b^2 - 4ac > 0$: a **real** value of $\sqrt{(b^2 - 4ac)}$ can be found, and so the equation has two real distinct roots;

  (ii) $b^2 - 4ac = 0$: the equation has two **real** identical roots;

 (iii) $b^2 - 4ac < 0$: there is no **real** value of $\sqrt{(b^2 - 4ac)}$ and so no **real** roots of the equation.

Exercise 1.1

Solve the following equations, stating whether the equation has two distinct real roots, two identical real roots or no real roots. If no real roots, leave the answer in the form

$$x = \frac{-b \pm \sqrt{(b^2 - 4ac)}}{2a}$$

N.B. You can use any suitable method, i.e. factorizing, completing the square, using the formula.

1. $9x^2 = 3x + 2$

2. $9x^2 = 8x + 2$

3. $9x^2 = 8x - 2$

4. $11x^2 = 48x$

5. $7x^2 - 38x + 15 = 0$

6. $9x^2 = 12x - 4$

Now we will explore further into the third possibility, when $b^2 - 4ac < 0$: i.e. when we need to define a new sort of number.

Definition: An **imaginary** number is one whose square is a negative real number; e.g.

$$\sqrt{-1}, \quad \sqrt{-7}, \quad \sqrt{-18}, \quad \sqrt{-25}$$

If $\sqrt{-1} = i$, then **all** other imaginary numbers can be expressed in terms of i.

(N.B. Sometimes the symbol j is used instead of i.)

Hence $\sqrt{-1} = i$; $\sqrt{-7} = i\sqrt{7}$; $\sqrt{-18} = 3\sqrt{2}\,i$; $\sqrt{-25} = 5i$.

Since $i = \sqrt{-1}$, then $i^2 = -1$

$$i^3 = i^2 i = -i$$

$$i^5 = i$$

$$\frac{1}{i^3} = \frac{i}{i^4} = \frac{i}{1} = i$$

Now we can solve equations like

$$x^2 - 2x + 2 = 0$$

so that $x = \dfrac{+2 \pm \sqrt{(4-8)}}{2} = \dfrac{2}{2} \pm \dfrac{\sqrt{-4}}{2} = 1 \pm \dfrac{2\sqrt{-1}}{2}$

and $x = 1 \pm i$

We have here a new sort of number with a real part and an imaginary part, and it is called a **complex** number.

Exercise 1.2

1. Simplify: i^9, i^{-3}, i^{-5}, i^{4n}, i^{4n+1}, i^{4n+3} when n is an integer, writing the answers as complex numbers.
2. Complete Exercise 1.1, Q3 by writing the two roots as complex numbers.

REMEMBER THAT

(i) **Definition**. A number in the form $a + bi$ where a and b are real and $i^2 = -1$, is called a **complex number**.

(ii) a or b or both can equal zero, as zero, 0 is a real number.

(iii) The field of complex numbers includes the set of real numbers and the set of imaginary numbers.
i.e. $-2 + 5i$, 3, $11i$, $1/2 + (\sqrt{3}/2)i$ are all complex numbers.

(iv) Two complex numbers are equal if and only if the real terms and the imaginary terms are both equal.

(v) Re $(a + bi) =$ real part of $a + bi = a$
Im $(a + bi) =$ imaginary part of $a + bi = bi$

(vi) It is frequently convenient to have a single letter to represent a complex number and the usual choice for this is z, and sometimes w. Thus $z = a + bi$, for example.

(vii) **Notation.** The standard symbol for the set of complex numbers is $\mathbb{C}$.

Exercise 1.1

Worked answers

Solve

1. $9x^2 = 3x + 2$

   $9x^2 - 3x - 2 = 0$ and this **will** factorize

   $(3x - 2)(3x + 1) = 0$

   Thus $x = \dfrac{2}{3},\ -\dfrac{1}{3}$

   i.e. two distinct real roots

   N.B. If you find factorizing hard, you **could** use the formula, but ultimately you **must** be able to factorize when possible at this level of work.

2. $9x^2 = 8x + 2$

   $9x^2 - 8x - 2 = 0$ which is not factorizable

   $$x = \frac{8 \pm \sqrt{64 + 72}}{18} = \frac{8 \pm \sqrt{136}}{18} = \frac{8 \pm 2\sqrt{34}}{18} = \frac{4 \pm \sqrt{34}}{9}$$

   thus $x = 1.092$ or -0.203 to 3 decimal places

   i.e. two distinct real roots.

3. $9x^2 = 8x - 2$

   $9x^2 - 8x + 2 = 0$ which does not factorize

   $$x = \frac{8 \pm \sqrt{64 - 72}}{18} = \frac{8 \pm \sqrt{-8}}{18} = \frac{8 \pm 2\sqrt{-2}}{18} = \frac{4 \pm \sqrt{-2}}{9}$$

   i.e. no real roots

4. $11x^2 = 48x$

   $11x^2 - 48x = 0$

   $x(11x - 48) = 0$

   $x = 0,\ \dfrac{48}{11}$, i.e. two distinct real roots

5. $7x^2 - 38x + 15 = 0$ and this will factorize

   $(7x - 3)(x - 5) = 0$

   $x = \dfrac{3}{7},\ 5$, i.e. two distinct real roots

6. $9x^2 = 12x - 4$

$9x^2 - 12x + 4 = 0$

$(3x - 2)(3x - 2) = 0$

$x = \dfrac{2}{3}, \; \dfrac{2}{3}, \;$ i.e. two identical real roots

Exercise 1.2

Worked answers

1. Simplify:

$$i^9 = i^8 \times i = 1 \times i = i$$

$$i^{-3} = \frac{1}{i^3} = \frac{i}{i^4} = \frac{i}{1} = i$$

$$i^{-5} = \frac{1}{i^5} = \frac{i}{i^6} = \frac{i}{-1} = -i$$

$$i^{4n} = (i^4)^n = 1^n = 1$$

$$i^{4n+1} = i^{4n} \times i = 1 \times i = i$$

$$i^{4n+3} = i^{4n} \times i^3 = 1 \times i^3 = -i \qquad (\text{i.e. } i^2 = -1)$$

2. The answer to Exercise 1.1, Q3, was

$$x = \frac{4 \pm \sqrt{-2}}{9}$$

and as complex numbers these answers are

$$x = \frac{4}{9} + \frac{\sqrt{2}}{9}i \text{ and } x = \frac{4}{9} - \frac{\sqrt{2}}{9}i$$

2 The algebra of complex numbers

This is easy, as, of course, algebra is algebra, nothing changes, you just remember that $i^2 = -1$.

Addition and subtraction: $z + w$ and $z - w$, where z and w are for example: $2 + 5i = z$ and $-3 + 7i = w$.

1. $(2 + 5i) + (-3 + 7i) = (2 - 3) + (5i + 7i) = -1 + 12i$
$$= z + w$$

2. $(2 + 5i) - (-3 + 7i) = (2 + 3) + (5i - 7i) = 5 - 2i$
$$= z - w$$

Multiplication: zw

3. $(2 + 5i)(-3 + 7i) = -6 - 15i + 14i + 35i^2$
$$= -6 - i - 35$$
$$= -41 - i = zw$$

Division: z/w

4. $\dfrac{2 + 5i}{-3 + 7i}$

Here, you must remember how you dealt with problems with surds like $\dfrac{3 + \sqrt{5}}{3 - \sqrt{5}}$ where you wanted to 'rationalize the denominator' (get rid of the surd on the bottom!). You learned to multiply top and bottom by the denominator with the sign in front of the surd term changed; e.g in this example

$$\frac{3 + \sqrt{5}}{3 - \sqrt{5}} = \frac{(3 + \sqrt{5})(3 + \sqrt{5})}{(3 - \sqrt{5})(3 + \sqrt{5})}$$

$$= \frac{9 + 3\sqrt{5} + 3\sqrt{5} + 5}{9 - 3\sqrt{5} + 3\sqrt{5} - 5}$$

$$= \frac{14 + 6\sqrt{5}}{4} = \frac{7}{2} + \frac{3\sqrt{5}}{2}$$

(Continued overleaf)

A similar technique works here —

$$\frac{2+5i}{-3+7i} = \frac{(2+5i)(-3-7i)}{(-3+7i)(-3-7i)}$$

$$= \frac{-6 - 15i - 14i - 35i^2}{9 - 21i + 21i - 49i^2}$$

$$= \frac{-6 + 35 - 29i}{9 + 49}$$

$$= \frac{29 - 29i}{58}$$

$$= \frac{1}{2} - \frac{1}{2}i$$

N.B. This cancelling is **luck**: the convenient numbers are **luck**: it is not always so obliging!

This technique, used for division, uses the **conjugate** of a complex number.

Definition: If $z = a + bi$ is any complex number, then its **conjugate**, denoted by $\bar{z}$ or z^* is $a - bi$.

Thus $4 + 3i$ and $4 - 3i$ are conjugate complex numbers, and $4 + 3i$ is said to be the conjugate of $4 - 3i$ and vice versa.

Exercise 2.1

1. If z_1 is $4 - i$ and z_2 is $3 + 2i$, find, in the form $a + bi$, where a and b are real:

   (i) $z_1 + z_2$ (ii) $z_1 - z_2$

   (iii) $z_1 z_2$ (iv) $\dfrac{z_1}{z_2}$

2. Given $z = -1 + 3i$, express $z + 2/z$, in the form $a + bi$, where a and b are real.

3. Solve the following equations for x and y.

   (i) $x + yi = (3 + i)(2 - 3i)$
   (ii) $x + yi = 3$
   (iii) $x + yi = 2i$
   (iv) $2 + 3i = (x + yi)(1 - i)$

4. Given that z and its conjugate $\bar{z}$ satisfy the equation:

$$z\bar{z} + 2iz = 12 + 6i$$

   find possible values of z.

There are three more things required in this chapter.
FIRST, to find square roots of a complex number.

If $z^2 = 3 + 4i$, find z, OR find the square roots of $3 + 4i$.
This is quite an easy problem, but many people don't know or can't remember how to get started.
Always, in this sort of question, let the unknown answer $= a + bi$.
Then the problem becomes easy.

$$(a + bi)^2 = 3 + 4i \qquad \text{thus to find } a \text{ and } b$$
$$a^2 - b^2 + 2abi = 3 + 4i$$
$$a^2 - b^2 = 3 \quad (1) \qquad \text{and } 2ab = 4 \quad (2)$$
$$\text{i.e. } ab = 2$$
$$a = \frac{2}{b}$$

and substituting in (1)

$$\left(\frac{2}{b}\right)^2 - b^2 = 3$$
$$4 - b^4 = 3b^2$$
$$b^4 + 3b^2 - 4 = 0$$
$$(b^2 + 4)(b^2 - 1) = 0$$

$b^2 = 1 \Rightarrow b \pm 1$ (remember the definition of a complex number says that a and b are real, thus $b^2 \neq -4$)

Now substitute in (2), when $b = 1$, $\quad a = 2$
$$\text{when } b = -1, \ a = -2$$

thus the square roots of $3 + 4i$ are $2 + i$ and $-2 - i$, i.e. $\pm(2 + i)$

N.B. You can always check if these are the correct answers by squaring them.

SECOND, complex numbers of a quadratic equation and factors of a similar expression.

Solve $x^2 + x + 1 = 0$ (1)

clearly this will not factorize and thus by the formula,

$$x = \frac{-1 \pm \sqrt{(1-4)}}{2} = -\frac{1}{2} \pm \frac{\sqrt{3}}{2}i$$

i.e. the two roots are

$$x = -\frac{1}{2} + \frac{\sqrt{3}}{2}i \text{ and } -\frac{1}{2} - \frac{\sqrt{3}}{2}i$$

and these are **conjugate** complex numbers.

*If one root of a quadratic equation with real coefficients is known and is complex, the other root is **always** the conjugate of the first.*

To factorize the expression $x^2 + x + 1$

$$= \left(x - \left(-\frac{1}{2} + \frac{\sqrt{3}}{2}i\right)\right)\left(x - \left(-\frac{1}{2} - \frac{\sqrt{3}}{2}i\right)\right)$$

i.e. $\left(x + \frac{1}{2} - \frac{\sqrt{3}}{2}i\right)\left(x + \frac{1}{2} + \frac{\sqrt{3}}{2}i\right)$

which is most easily obtained from roots of equation (1) above.

Exercise 2.2

1. Find the two square roots of
   (i) $5 - 12i$
   (ii) $8i$
   (iii) $-7 + 24i$

2. Solve the equations
   (i) $x^2 + 4x + 5 = 0$
   (ii) $x^2 - 2x + 17 = 0$

3. Factorize the expressions
   (i) $x^2 + 4x + 5$
   (ii) $x^2 - 2x + 17$
   (iii) $x^3 + x^2 - 2$ using factor theorem to find the first real factor.

THIRD, to find the roots of $z^3 = 1$.
Obviously, one root is $z = 1$.

$$z^3 - 1 = 0 \Rightarrow (z-1)(z^2 + z + 1) = 0$$

and solving $z^2 + z + 1 = 0$ gives $z = \dfrac{-1 \pm \sqrt{(1-4)}}{2} = -\dfrac{1}{2} \pm \dfrac{\sqrt{3}}{2}i$

Thus $z^3 = 1$ gives $z = 1, \ -\dfrac{1}{2} + \dfrac{\sqrt{3}}{2}i, \ -\dfrac{1}{2} - \dfrac{\sqrt{3}}{2}i$.

This is a very important result and will be developed and explored further in Chapters 10–20.
For now, think about how you would find the roots of $z^7 = 1$, for example. Clearly the above method cannot help.
Why must it always be '1'? What about $z^4 = -1$? What about $z^5 = 2$? What about $z^8 = 1 + i$?

All this is studied in Chapters 10–20.
For now, just be aware of these problems.

Exercise 2.1

Worked answers

1. $z_1 = 4 - i$ and $z_2 = 3 + 2i$

   (i) $\therefore z_1 + z_2 = (4 - i) + (3 + 2i) = 7 + i$

   (ii) $z_1 - z_2 = (4 - i) - (3 + 2i) = 1 - 3i$

   (iii) $z_1 z_2 = (4 - i)(3 + 2i) = 12 - 3i + 8i - 2i^2$

   $$= 12 + 5i + 2 = 14 + 5i$$

   (iv) $\dfrac{z_1}{z_2} = \dfrac{4 - i}{3 + 2i} = \dfrac{(4 - i)(3 - 2i)}{(3 + 2i)(3 - 2i)} = \dfrac{12 - 11i + 2i^2}{9 - 4i^2}$

   $$= \frac{10}{13} - \frac{11}{13}i$$

2. $z = -1 + 3i, \ \therefore z + \dfrac{2}{z} = (-1 + 3i) + \dfrac{2(-1 - 3i)}{(-1 + 3i)(-1 - 3i)}$

   $$= (-1 + 3i)\frac{-2 - 6i}{1 - 9i^2}$$

   $$= (-1 + 3i) - \frac{\cancel{2}^1}{\cancel{10}5} - \frac{\cancel{6}^3}{\cancel{10}5}i$$

   $$= -\frac{6}{5} + \frac{12}{5}i$$

3. (i) $x + yi = (3 + i)(2 - 3i) = 6 - 7i - 3i^2$

   $$= 9 - 7i \qquad\qquad \therefore x = 9, \ y = -7$$

   (ii) $x + yi = 3$ $\qquad\qquad\qquad\qquad \therefore x = 3, \ y = 0$

   (iii) $x + yi = 2i$ $\qquad\qquad\qquad\quad \therefore x = 0, \ y = 2$

   (iv) $2 + 3i = (x + yi)(1 - i)$

   $$\therefore x + yi = \frac{2 + 3i}{1 - i} = \frac{(2 + 3i)(1 + i)}{(1 - i)(1 + i)}$$

   $$= \frac{-1 + 5i}{1 + 1} = -\frac{1}{2} + \frac{5}{2}i \qquad \therefore x = -\frac{1}{2}, \ y = \frac{5}{2}$$

4. Let $z = a + bi$ $\therefore \bar{z} = a - bi$

$\qquad z\bar{z} + 2iz = 12 + 6i$

$\therefore (a + bi)(a - bi) + 2i(a + bi) = 12 + 6i$

$\qquad\qquad a^2 + b^2 + 2ai - 2b = 12 + 6i$

$\therefore a^2 + b^2 - 2b = 12$ (1)   and $2ai = 6i$ (2)

$\therefore$ from (2), $a = 3$ and substituting in (1) $9 + b^2 - 2b = 12$

$\qquad\qquad\qquad\qquad\qquad\qquad\qquad\qquad b^2 - 2b - 3 = 0$

$\qquad\qquad\qquad\qquad\qquad\qquad\qquad (b - 3)(b + 1) = 0$

$\therefore b = 3$ or -1.

$\therefore z = 3 + 3i$ or $z = 3 - i$

Exercise 2.2

Worked answers

1 (i) Let the square roots be $a + bi$ i.e. $a + bi = \pm\sqrt{(5 - 12i)}$

$\qquad\qquad (a + bi)^2 = 5 - 12i$

$\qquad a^2 - b^2 + 2abi = 5 - 12i$

$\qquad a^2 - b^2 = 5$ (1)   and $2ab = -12$ (2)

$\qquad\qquad\qquad\qquad$ so that $b = -\dfrac{6}{a}$

then $a^2 - \dfrac{36}{a^2} = 5$

$a^4 - 5a^2 - 36 = 0$

$(a^2 - 9)(a^2 + 4) = 0 \Rightarrow a^2 = 9$ because a^2 is real and $a^2 \neq -4$

$\qquad\qquad\qquad$ thus $a = \pm3$

when $a = 3$, $b = -2$ or when $a = -3$, $b = 2$

Thus square roots are $3 - 2i$ and $-3 + 2i$ i.e. $\pm(3 - 2i)$

(ii) Let square roots be $a + bi$ i.e. $a + bi = \pm\sqrt{8i}$

$\quad \therefore (a + bi)^2 = 8i$

i.e. $a^2 - b^2 = 0$ and $2ab = 8 \Rightarrow b = \dfrac{4}{a}$

$a^2 - \dfrac{16}{a^2} = 0 \Rightarrow a^4 = 16 \Rightarrow a = \pm2$

when $a = 2$, $b = 2$ and when $a = -2$, $b = -2$

square roots are $2 + 2i$ and $-2 - 2i$ i.e. $\pm(2 + 2i)$

(iii) Let square roots be $a + bi$ i.e. $a + bi = \pm\sqrt{(-7 + 24i)}$

$(a + bi)^2 = -7 + 24i$

$a^2 - b^2 = -7$ and $2ab = 24$

so that $b = \dfrac{12}{a}$

$a^2 - \dfrac{144}{a^2} = -7$

$a^4 - 7a^2 - 144 = 0$

$(a^2 - 16)(a^2 + 9) = 0$

$a^2 = 16$ because $a^2 \neq -9$ because a is real.

$a = \pm 4$ and thus $b = \pm 3$

Thus square roots are $4 + 3i$, $-4 - 3i$ i.e. $\pm(4 + 3i)$

Look harder here.

Have you checked each part by squaring the answers to see if they are correct as recommended earlier.

If you do this you will find that the answer to (iii) is wrong! It is easy to make a slip and therefore get a wrong answer but you can find most errors with a bit of care.

Right; back to part (iii) to put it right.

The error occurs on line 6, from line 5; $a^2 - 144/a^2 = -7$; which is right, to line 6; $a^4 - 7a^2 - 144 = 0$; which is *wrong.*

It should be $a^4 + 7a^2 - 144 = 0$

$(a^2 + 16)(a^2 - 9) = 0$

thus $a^2 = 9$ because $a^2 \neq -16$ because a is real.

and so $a = \pm 3$

when $a = 3$, $b = 4$ and when $a = -3$, $b = -4$

Thus square roots are $3 + 4i$ and $-3 - 4i$ i.e. $\pm(3 + 4i)$

now, check by squaring—do this check on a rough piece of paper—don't hand it in, there will be no marks for doing it, but you will LOSE marks if your answer is wrong as it was on page 131.

2 (i) To solve $x^2 + 4x + 5 = 0$

thus $x = \dfrac{-4 \pm \sqrt{16 - 20}}{2} = -2 \pm i$

(ii) To solve $x^2 - 2x + 17 = 0$

thus $x = \dfrac{2 \pm \sqrt{4 - 68}}{2} = \dfrac{2 \pm \sqrt{-64}}{2} = \dfrac{2 \pm 8i}{2} = 1 \pm 4i$

3 (i) Factorize $x^2 + 4x + 5$

$$= (x - (-2 + i))(x - (-2 - i))$$
$$= (x + 2 - i)(x + 2 + i) \qquad \text{(using 2(i))}$$

(ii) Factorize $x^2 - 2x + 17$

$$= (x - (1 + 4i))(x - (1 - 4i))$$
$$= (x - 1 - 4i)(x - 1 + 4i) \qquad \text{(using 2(ii))}$$

(iii) Factorize $x^3 + x^2 - 2 = f(x)$ say
$f(1) = 1 + 1 - 2 = 0$ thus one factor is $(x - 1)$ using factor theorem
$$f(x) = (x - 1)(x^2 + 2x + 2)$$

and considering $x^2 + 2x + 2 = 0$

$$x = \frac{-2 \pm \sqrt{4 - 8}}{2} = -1 \pm i$$

Thus $x^2 + 2x + 2 = (x - (-1 + i))(x^2 - (-1 - i))$
$$= (x + 1 - i)(x + 1 + i)$$

so that $f(x) = x^3 + x^2 - 2$
$$= (x - 1)(x^2 + 2x + 2)$$
$$= (x - 1)(x + 1 - i)(x + 1 + i)$$

3 The Argand diagram

J.R. Argand was an 18th century Swiss mathematician, who was one of the first to develop the visual representation of the complex number $a + bi$, using Cartesian coordinates of a point A say, where A is the point (a,b), and the vector $\overrightarrow{OA}$ is its visual representation on the diagram.

Thus this diagram has come to be known as the Argand diagram. Real numbers are represented on the x-axis – often therefore known as the real axis – and imaginary numbers on the y-axis – the imaginary axis.

A general complex number $x + yi$ is represented by the line $\overrightarrow{OP}$, where P is the point (x,y) and if $x + iy = z$, then z is shown on the diagram.

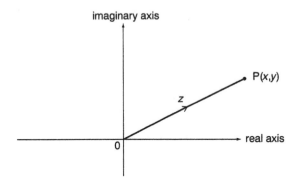

Thus **any** complex number can be represented by a point in the 2-dimensional
Cartesian plane; e.g.

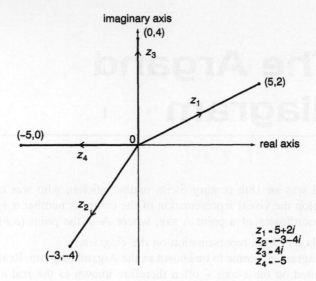

$z_1 = 5+2i$
$z_2 = -3-4i$
$z_3 = 4i$
$z_4 = -5$

4 The modulus, argument form for a complex number

Instead of representing the general complex number $x + iy$ by the Cartesian coordinates (x,y), there is another way of representing the point with polar coordinates (r,θ). If you have not yet studied polar coordinates you will still be able to deal with this very important work which is **absolutely necessary** for us to proceed with complex number theory.

Consider the complex number $x + iy$, represented on the Argand diagram below, by the point P(x,y), using Cartesian coordinates.

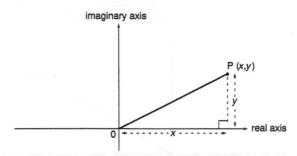

This same point P can be located in a very different way, by using its distance r from a point 0 (the origin), and the angle θ made by the line OP with the real axis.

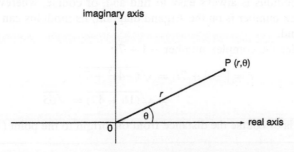

Hence P becomes the point (r,θ) and r and θ can easily be linked to the more familiar x and y:

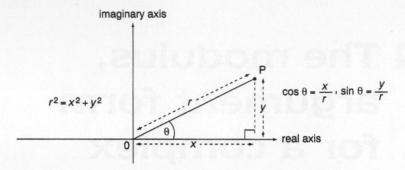

Therefore the general complex number $x + iy = r\cos\theta + ir\sin\theta$; i.e.

$$x + iy = r(\cos\theta + i\sin\theta)$$

and this is another form of the general complex number, which seems at the moment unnecessarily elaborate! Surely the $2 + 3i$ form is easier and will do all we want of it! The answer to this is **NO IT WON'T**, so you must persevere so that you can convert **ANY** complex number into its r,θ form, and will begin to see the advantages.

Definition: The length of OA is called the **modulus** of the complex number $a + bi$ and is written $|a + bi|$

$$r = \sqrt{(a^2 + b^2)} = |a + bi|$$

This modulus is always easy to find and, of course, wherever the complex number is on the Argand diagram, its modulus can easily be found.

Consider the complex number $-4 + 7i$:

$$r = |-4 + 7i| = \sqrt{(-4)^2 + 7^2}$$

$$= \sqrt{(16 + 49)} = \sqrt{65}$$

which is of course the distance from the origin to the point $(-4, 7)$.

N.B. A very common error here is to feel compelled to square the 'i' in the complex number! This means you have not really thought properly about the Argand diagram and I suggest you need to read pages 17 and 18 again and concentrate a bit harder!

Exercise 4.1

1. Write down the moduli (plural of modulus) of the following complex numbers.

   (i)　$3 + 4i$　　　　　(ii)　$-1 + i$

   (iii)　i　　　　　　(iv)　$-2 - 3i$

   (v)　-5　　　　　(vi)　$-\dfrac{1}{2} - \dfrac{\sqrt{3}}{2}i$

   (vii)　$-4 + 3i$

Definition: The angle θ, called the **argument** of the complex number $a + bi$, is written $\mathrm{Arg}(a + bi) = \theta = \arctan b/a$.

There are problems here, as there are an infinite number of angles that fit the definition.

Consider the number $1 + i$. Sketch an Argand diagram and look at θ.

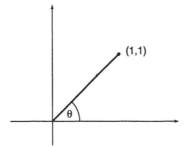

The obvious value for $\theta = \mathrm{Arg}(1 + i) = \pi/4$.

(I use radians unless degrees are specifically asked for – both are correct.)

Another value is $\theta = -7\pi/4$ and another, $9\pi/4$.
They are all correct for $\mathrm{Arg}(1 + i)$.

Obviously the definition needs to be tightened so that everyone gives the same answer, and this is called the principal value of the argument and is written $\arg(a + bi)$, where $-\pi < \theta \leq \pi$:

$$\arg(1 + i) = \frac{\pi}{4}$$

N.B. $\arg(-1-i) = -3\pi/4$. *Always sketch the point on the Argand diagram to make sure that you are writing sense!*

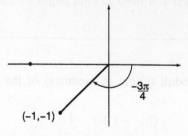

N.B. The **argument** *is sometimes given an alternative name,* **amplitude**.

Here are some examples to help you:
Consider the complex numbers

(i) $z_1 = -1 - i$ (ii) $-5 + 12i = z_2$ (iii) $2 - i = z_3$

   (a) Find the modulus of each.
   (b) Find the argument of each (principal value).
   (c) Express each in the form $r(\cos \theta + i \sin \theta)$.

N.B. Sketch each point on the Argand diagram to enable you to check the answers, especially the position and size of the argument.

(i) $z_1 = 1 - i$

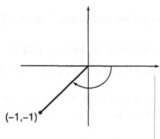

$(-1,-1)$

   (a) $|-1 - i| = \sqrt{((-1)^2 + (-1)^2)} = \sqrt{2}.$

   (b) $\arg(-1 - i) = -\dfrac{3\pi}{4}$

   (c) $z_1 = \sqrt{2}\left(\cos\left(-\dfrac{3\pi}{4}\right) + i \sin\left(-\dfrac{3\pi}{4}\right)\right)$

This is easily done by looking at the diagram, where $|z_1|$ is obviously $\sqrt{2}$, and $\arg z_1 = -3\pi/4$ without any complicated calculations.

(ii) $z_2 = -5 + 12i$

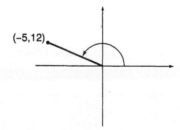

$(-5,12)$

   (a) $|-5 + 12i| = 13$

   (b) $\arg z_2 = \text{positive and obtuse} = \arctan\left(-\dfrac{12}{5}\right)$

         $= 1.966^c$ to 3 decimal places

   (c) $z_2 = 13(\cos 1.966^c + i \sin 1.966^c)$

(iii) $z_3 = 2 - i$

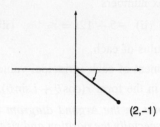

(2,−1)

(a) $|z_3| = \sqrt{5}$

(b) $\arg z_3 = $ negative and acute $= \arctan\left(-\dfrac{1}{2}\right)$

$= -0.464$ to 3 decimal places

(c) $z_3 = \sqrt{5}(\cos(-0.464^c) + i\sin(-0.464^c))$

As you can see, the modulus is never very difficult to find, and the argument, once you have practised with the calculator, is also not so difficult.

There is another way you may prefer to do these, so let me show you this alternative.

Same examples as on the previous page.

(i) $z_1 = -1 - i$

   (a) Find the modulus as before and sketch the diagram as before.

      $|z_1| = \sqrt{2}.$

   (b) and (c) Then write $z_1 = \sqrt{2} - \dfrac{1}{\sqrt{2}} - \dfrac{1}{\sqrt{2}}i$ and find θ such that $\cos\theta = -1/\sqrt{2}$ and $\sin\theta = -1/\sqrt{2}.$

Remember you already know from the diagram that the point is in the 3rd quadrant. You should be able to work out that $45°$ or $\pi/4$ is relevant here and therefore, from the diagram, the argument must be $-3\pi/4$.

(ii) $z_2 = -5 + 12i$

   $|z_2| = 13$

   thus $z_2 = 13 - \dfrac{5}{13} + \dfrac{12}{13}i$ where $\cos\theta = -5/13$ and $\sin\theta = 12/13$, where θ is in the 2nd quadrant, from the diagram.

   As before, a calculator will be needed to find θ.

   Similarly for (iii) $z_3 = 2 - i$.

I prefer this method for non-calculator questions, i.e. those involving $30°$, $45°$, $60°$ and larger similar angles, i.e. $300°$, $630°$, or the equivalents in radians. You must of course know the trigonometric values for these angles very well indeed.

$$-2 + 2i = \sqrt{8}\left(-\frac{2}{2\sqrt{2}} + \frac{2}{2\sqrt{2}}i\right) = 2\sqrt{2}\left(-\frac{1}{\sqrt{2}} + \frac{1}{\sqrt{2}}i\right)$$

$$= 2\sqrt{2}\left(\cos\frac{3\pi}{4} + i\sin\frac{3\pi}{4}\right)$$

Always draw a diagram to help locate the angle. In this example it is obvious that the argument is $3\pi/4$ or $135°$.

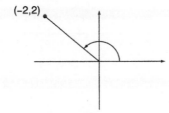

Now you try Exercise 4.2. Don't look at my worked answers until you have **really** tried.

Exercise 4.2

1. Find the modulus and the principal value of the argument of:

   (i) $z_1 = 1 + i$ (ii) $z_2 = 1 + \sqrt{3}i$

   (iii) $z_3 = -1 + \sqrt{3}i$ (iv) $z_4 = -2$

   (v) $z_5 = 3i$

N.B. Do not use a calculator for any of these.

2. Now write each of question 1's numbers in the form

$$r(\cos\theta + i\sin\theta)$$

3. Express each of the following in the form $r(\cos\theta + i\sin\theta)$, giving θ in both radians to 3 significant figures and degrees to the nearest degree.

   (i) $z_6 = 24 + 7i$ (ii) $z_7 = -24 + 7i$

4. Given $|z| = 20$ and $\arg z = 120°$, find z in the form $a + bi$.

5. Find the principal values of the arguments of:

   (i) $-i$ (ii) $\cos\dfrac{7\pi}{3} + i\sin\dfrac{7\pi}{3}$

   (iii) $\cos\dfrac{2\pi}{3} - i\sin\dfrac{2\pi}{3}$

Exercise 4.1

Worked answers

1 (i) $|3 + 4i| = \sqrt{(3^2 + 4^2)} = 5$

(ii) $|-1 + i| = \sqrt{((-1)^2 + 1^2)} = \sqrt{2}$

(iii) $|i| = 1$

(iv) $|-2 - 3i| = \sqrt{((-2)^2 + (-3)^2)} = \sqrt{13}$

(v) $|-5| = 5$

(vi) $\left| -\dfrac{1}{2} - \dfrac{\sqrt{3}}{2}i \right| = \sqrt{\left(\left(-\dfrac{1}{2} \right)^2 + \left(-\dfrac{\sqrt{3}}{2} \right)^2 \right)} = \sqrt{\left(\dfrac{1}{4} + \dfrac{3}{4} \right)} = 1$

(vii) $|-4 + 3i| = \sqrt{((-4)^2 + 3^2)} = \sqrt{(16 + 9)} = 5$

Exercise 4.2

Worked answers

1 (i) $z_1 = 1 + i$: First, look at z_1 on the Argand diagram.

   $|z_1| = \sqrt{2}$ $\arg z_1 = \dfrac{\pi}{4}$ (or $45°$)   i.e. 1st quadrant

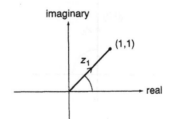

(ii) $z_2 = 1 + \sqrt{3}i$: Again, look at z_2 on the Argand diagram.

   $|z_2| = \sqrt{(1^2 + \sqrt{3}^2)} = \sqrt{(1 + 3)} = 2$

   $\arg z_2 = \dfrac{\pi}{3}$ (or $60°$ — 1st quadrant)

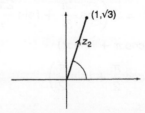

(iii) $z_3 = -1 + \sqrt{3}i$: Use common sense and the knowledge you gained in (ii)

$$|z_3| = 2 \qquad \arg z_3 = 2\pi/3 \text{ (or } 120° - \text{ 2nd quadrant)}$$

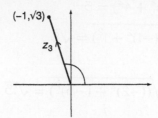

(iv) $z_4 = -2$: Use common sense.

$$|z_4| = 2 \qquad \arg z_4 = \pi \text{ (or } 180°)$$

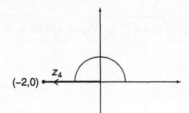

(v) $z_5 = 3i$: Use common sense.

$$|z_5| = 3 \qquad \arg z_5 = \frac{\pi}{2} \text{ (or } 90°)$$

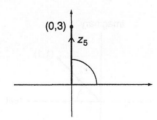

2. From Q1.

   (i) $z_1 = 1 + i = \sqrt{2}\left(\cos\dfrac{\pi}{4} + i\sin\dfrac{\pi}{4}\right)$

   (ii) $z_2 = 1 + \sqrt{3}i = 2\left(\cos\dfrac{\pi}{3} + i\sin\dfrac{\pi}{3}\right)$

   (iii) $z_3 = -1 + \sqrt{3}i = 2\left(\cos\dfrac{2\pi}{3} + i\sin\dfrac{2\pi}{3}\right)$

   (iv) $z_4 = -2 = 2(\cos\pi + i\sin\pi)$

   (v) $z_5 = 3i = 3\left(\cos\dfrac{\pi}{2} + i\sin\dfrac{\pi}{2}\right)$

3. Here a calculator will be needed, because of asking for the answer to 2 significant figures etc.

(i) $z_6 = 24 + 7i$ (1st quadrant),

$$|z_6| = \sqrt{(24^2 + 7^2)} = 25.$$

$$\arg z_6 = \arctan \frac{7}{24} = 0.28^c \text{ to 2 significant figures}$$

$$= 16° \text{ to nearest degree.}$$

$$z_6 = 25(\cos 0.28^c + i \sin 0.28^c) = 25(\cos 16° + i \sin 16°)$$

(ii) $z_7 = -24 + 7i$: here common sense and knowledge from the last question will help.

$$|z_7| = 25 \text{ (2nd quadrant)}$$

$$\arg z_7 = \pi - \arctan \frac{7}{24} = 2.9^c \text{ to 2 significant figures}$$

$$= 164° \text{ to nearest degree}$$

$$z_7 = 25(\cos 2.9^c + i \sin 2.9^c) = 25(\cos 164° + i \sin 164°)$$

4. $|z| = 20$, $\arg z = 120°$ thus

$$z = 20(\cos 120° + i \sin 120°)$$

$$= 20 \left(-\frac{1}{2} + i \frac{\sqrt{3}}{2} \right) = -10 + 10\sqrt{3}i$$

N.B. Q3 and Q4: you **could** have sketched Argand diagrams as in Q1. There will be no extra marks unless the diagram has been specifically asked for, but they will always act as a check for you, and will draw your attention to any careless mistake you might have made. You could also try visualizing the relevant Argand diagram without bothering to draw it.

5. To find the principal values of the arguments of

(i) $z = -i$.

(1) Think of its position on the Argand diagram.

(2) Sketch it if you want to, but it shouldn't be necessary by now, if you have done the earlier chapters properly.

(3) $\arg z = -\pi/2$ or $\arg z = -90°$.

(ii) $z = \cos(7\pi/3) + i \sin(7\pi/3)$. Thus $\text{Arg } z = 7\pi/3$, but this is **not** the principal value of the argument because $7\pi/3$ is **not** in the correct range: i.e. $-\pi < \theta \leq \pi$. So, think of its position on the Argand diagram.

$\arg z = \pi/3$ (i.e. $(7\pi/3) - 2\pi$) and $\pi/3$ is in required range.

(iii) $z = \cos(2\pi/3) - i\sin(2\pi/3)$.

This is **not** the required form—i.e. $z = \cos\theta + i\sin\theta$.

This can **always** be achieved by replacing the angle, in this case $2\pi/3$, by $-2\pi/3$, because $\cos 2\pi/3 = \cos(-2\pi/3)$ and $\sin(2\pi/3) = -\sin(-2\pi/3)$

thus $z = \cos(-2\pi/3) + i\sin(-2\pi/3)$

therefore $\arg z = -2\pi/3$

5 Products and quotients, using the $r(\cos\theta + i\sin\theta)$ form

An interesting and useful thing happens when two complex numbers in the r,θ form are multiplied together.

Let $z_1 = r_1(\cos\theta_1 + i\sin\theta_1)$ and $z_2 = r_2(\cos\theta_2 + i\sin\theta_2)$

Therefore
$$z_1 z_2 = r_1 r_2(\cos\theta_1 + i\sin\theta_1)(\cos\theta_2 + i\sin\theta_2)$$
$$= r_1 r_2(\cos\theta_1 \cos\theta_2 + i^2\sin\theta_1 \sin\theta_2 + i\sin\theta_1 \cos\theta_2 + i\cos\theta_1 \sin\theta_2)$$
$$= r_1 r_2((\cos\theta_1 \cos\theta_2 - \sin\theta_1 \sin\theta_2) + i(\sin\theta_1 \cos\theta_2 + \cos\theta_1 \sin\theta_2))$$

Using the trig formulae
$$\cos(A + B) = \cos A \cos B - \sin A \sin B$$
and
$$\sin(A + B) = \sin A \cos B - \cos A \sin B$$
so that
$$z_1 z_2 = r_1 r_2(\cos(\theta_1 + \theta_2) + i(\sin(\theta_1 + \theta_2)))$$
$$= R(\cos\theta + i\sin\theta)$$

which is in the correct r,θ form.

Thus when two complex numbers in the r,θ form are multiplied together, the result is **ALSO** a complex number in the same form where:
(1) the modulus of the products is the product of the moduli; i.e.

$$|z_1 z_2| = |z_1||z_2|$$

and (2) the argument of the product is the sum of the arguments (not necessarily the principal argument); i.e.

$$\text{Arg}\, z_1 z_2 = \text{Arg}\, z_1 + \text{Arg}\, z_2$$

*N.B. The above theory is sometimes asked for as part of an A Level question. It is tedious to write out, but **make sure** that you **CAN** do it, as it will obviously earn you marks.*

Otherwise, the result is very useful for answering questions like this, which are very common in A Level papers.

Find the modulus and argument of z_1z_2, where $z_1 = 2 + 2i$ and $z_2 = 1 + \sqrt{3}i$, and hence write z_1z_2 in the form $r(\cos\theta + i\sin\theta)$. *In my opinion, it is more easily done using the previous page*; i.e.

$$z_1 = 2 + 2i = 2\sqrt{2}\left(\frac{1}{\sqrt{2}} + \frac{1}{\sqrt{2}}i\right) = 2\sqrt{2}\left(\cos\frac{\pi}{4} + i\sin\frac{\pi}{4}\right)$$

$$z_2 = 1 + \sqrt{3}i = 2\left(\frac{1}{2} + \frac{\sqrt{3}}{2}i\right) = 2\left(\cos\frac{\pi}{3} + i\sin\frac{\pi}{3}\right)$$

thus $|z_1z_2| = |z_1||z_2| = 2\sqrt{2}\cdot 2 = 4\sqrt{2}$

$$\text{Arg}\,z_1z_2 = \text{Arg}\,z_1 + \text{Arg}\,z_2 = \frac{\pi}{4} + \frac{\pi}{3} = \frac{7\pi}{12}$$

thus $z_1z_2 = 4\sqrt{2}\left(\cos\frac{7\pi}{12} + i\sin\frac{7\pi}{12}\right)$

If you prefer, find z_1z_2 first: i.e.

$$z_1z_2 = (2 + 2i)(1 + \sqrt{3}i) = 2 - 2\sqrt{3} + (2 + 2\sqrt{3})i$$

Now to find $|z_1z_2|$, you must work out $\sqrt{((2-2\sqrt{3})^2+(2+2\sqrt{3})^2)}$ —if you are good at basic algebra this is not too difficult, and it will eventually come to $4\sqrt{2}$, but surely it is less taxing to do it as above.

Now $\text{Arg}\,z_1z_2$. This is the second quadrant and so will be

$$\pi - \arctan\frac{2 + 2\sqrt{3}}{2 - 2\sqrt{3}}$$

and again will come to the same answer as above, but with much more difficulty.

Similarly, with z_1 and z_2 as on the previous page,

$$\frac{z_1}{z_2} = \frac{r_1}{r_2}(\cos(\theta_1 - \theta_2) + i\sin(\theta_1 - \theta_2))$$

i.e. $\left|\frac{z_1}{z_2}\right| = \frac{|z_1|}{|z_2|}$ and $\text{Arg}\frac{z_1}{z_2} = \text{Arg}\,z_1 - \text{Arg}\,z_2$.

N.B. Therefore a complex equation of the type $\left|\frac{z - 3}{z + 5i}\right| = 3$ *can be transformed into* $|z - 3| = 3|z + 5i|$ *which is much more easily solved.*

Now try Exercise 5.1.

Exercise 5.1

1. If $z_1 = \dfrac{1}{2} + \dfrac{\sqrt{3}}{2}i$ and $z_2 = \dfrac{\sqrt{3}}{2} + \dfrac{1}{2}i$, convert z_1 and z_2 into the r, θ form and hence find $z_1 z_2$ and (z_1/z_2).

2. Hence similarly when
   (i) $z_1 = 1 - \sqrt{3}i$ and $z_2 = 1 - i$.
   (ii) $z_1 = z_2 = 1 + i$.
   (iii) $z_1 = -1 + i$ and $z_2 = -1 - \sqrt{3}i$.

3. London. A Level. Core Syllabus
   Given that $z_1 = -i$ and $z_2 = 1 + i\sqrt{3}$, find the modulus and argument of:

   (a) $z_1 z_2$ (b) $\dfrac{z_1}{z_2}$

4. London. A Level. Core Syllabus
   Find the modulus and argument of each of the complex numbers z_1 and z_2, where $z_1 = 1 + i$, $z_2 = \sqrt{3} - i$.

   Hence, or otherwise, show that $\arg\left(\dfrac{z_1}{z_2}\right) = \dfrac{5\pi}{12}$.

Exercise 5.1

Worked answers

1. $z_1 = \dfrac{1}{2} + \dfrac{\sqrt{3}}{2}i = \cos 60° + i\sin 60°$ N.B. $|z_1| = 1$

(using degrees)

$z_2 = \dfrac{\sqrt{3}}{2} + \dfrac{1}{2}i = \cos 30° + i\sin 30°$ $|z_2| = 1$

thus $z_1 z_2 = \cos(60° + 30°) + i\sin(60° + 30°)$

$= \cos 90° + i\sin 90° = i$

$\dfrac{z_1}{z_2} = \cos(60° - 30°) + i\sin(60° - 30°)$

$= \cos 30° + i\sin 30° = \dfrac{\sqrt{3}}{2} + \dfrac{1}{2}i$

2. (i) $z_1 = 1 - \sqrt{3}i = 2\left(\dfrac{1}{2} - \dfrac{\sqrt{3}}{2}i\right) = 2(\cos(-60°) + i\sin(-60°))$

(using degrees)

$z_2 = 1 - i = \sqrt{2}\left(\dfrac{1}{\sqrt{2}} - \dfrac{1}{\sqrt{2}}i\right) = \sqrt{2}(\cos(-45°) + i\sin(-45°))$

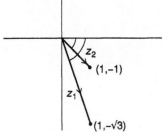

thus $z_1 z_2 = 2\sqrt{2}(\cos(-60° - 45°) + i\sin(-60° - 45°))$

$= 2\sqrt{2}(\cos(-105°) + i\sin(-105°))$

$\dfrac{z_1}{z_2} = \dfrac{2}{\sqrt{2}}(\cos(-60° + 45°) + i\sin(-60° + 45°))$

$= \sqrt{2}\cos(-15°) + i\sin(-15°))$

(ii) $z_1 = z_2 = 1 + i = \sqrt{2}(\cos 45° + i\sin 45°)$ (using degrees)

thus $z_1 z_2 = z^2 = \sqrt{2} \cdot \sqrt{2}(\cos(45° + 45°) + i\sin(45° + 45°)$
$= 2(\cos 90° + i\sin 90°) = 2i$

$\dfrac{z_1}{z_2} = 1$ (common sense, but of course the theory still works)

(iii) $z_1 = -1 + i = \sqrt{2}\left(-\dfrac{1}{\sqrt{2}} + \dfrac{1}{\sqrt{2}}i\right)$

$= \sqrt{2}\left(\cos\dfrac{3\pi}{4} + i\sin\dfrac{3\pi}{4}\right)$ (using radians)

$z_2 = -1 - \sqrt{3}i = 2\left(-\dfrac{1}{2} - \dfrac{\sqrt{3}}{2}i\right)$

$= 2\left(\cos\left(-\dfrac{2\pi}{3}\right) + i\sin\left(-\dfrac{2\pi}{3}\right)\right)$

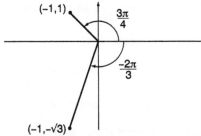

thus $z_1 z_2 = 2\sqrt{2}\left(\cos\left(\dfrac{3\pi}{4} - \dfrac{2\pi}{3}\right) + i\sin\left(\dfrac{3\pi}{4} - \dfrac{2\pi}{3}\right)\right)$

$= 2\sqrt{2}\left(\cos\dfrac{\pi}{12} + i\sin\dfrac{\pi}{12}\right)$

$\dfrac{z_1}{z_2} = \dfrac{\sqrt{2}}{2}\left(\cos\left(\dfrac{3\pi}{4} + \dfrac{2\pi}{3}\right) + i\sin\left(\dfrac{3\pi}{4} + \dfrac{2\pi}{3}\right)\right)$

$= \dfrac{\sqrt{2}}{2}\left(\cos\dfrac{17\pi}{12} + i\sin\dfrac{17\pi}{12}\right)$

N.B. (1) If units have **not** been stipulated, you can work in radians or degrees — but **don't** use a mixture.

(2) The answer does **not** have to give the **principal** argument, unless it is stipulated; e.g. 2(iii) z_1/z_2. Here the argument is **not** the principal value.

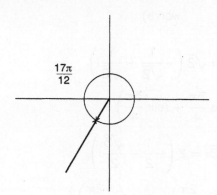

If it **is** required, use the Argand diagram.

You can see that $\text{Arg}\,\dfrac{z_1}{z_2} = \dfrac{17\pi}{12}$ so that $\arg\dfrac{z_1}{z_2} = -\dfrac{7\pi}{12}$

3. London. A Level. Core Syllabus.

$$z_1 = -i \qquad |z_1| = 1 \qquad \arg z_1 = -\frac{\pi}{2}$$

$$z_2 = 1 + i\sqrt{3} \qquad |z_2| = 2 \qquad \arg z_2 = \frac{\pi}{3}$$

(N.B. you can work in degrees if you prefer)

(a) $|z_1 z_2| = |z_1||z_2| = 2$

$$\arg z_1 z_2 = \arg z_1 + \arg z_2 = -\frac{\pi}{2} + \frac{\pi}{3} = -\frac{\pi}{6}$$

(b) $\left|\dfrac{z_1}{z_2}\right| = \dfrac{|z_1|}{|z_2|} = \dfrac{1}{2}$

$$\arg \frac{z_1}{z_2} = \arg z_1 - \arg z_2 = -\frac{\pi}{2} - \frac{\pi}{3} = -\frac{5\pi}{6}$$

OR

$z_1 = -i,\ z_2 = 1 + i\sqrt{3}$

so that $z_1z_2 = -i(1 + i\sqrt{3}) = -i - i^2\sqrt{3} = \sqrt{3} - i$

thus $|z_1z_2| = 2$ and $\arg z_1z_2 = -\dfrac{\pi}{6}$

$$\frac{z_1}{z_2} = \frac{-i}{1 + i\sqrt{3}} = \frac{-i(1 - i\sqrt{3})}{(1 + i\sqrt{3})(1 - i\sqrt{3})} = \frac{-i - \sqrt{3}}{4}$$

$$= \frac{-\sqrt{3}}{4} - \frac{i}{4} \qquad *$$

$$\left|\frac{z_1}{z_2}\right| = \frac{1}{2},\ \arg\frac{z_1}{z_2} = -\frac{5\pi}{6}$$

N.B. This may seem easier to you, **BUT** look at $-\dfrac{\sqrt{3}}{4} - \dfrac{i}{4}$

(1) You must be an accurate, careful worker to get this result — look at line *

(2) Can you work out $\left|\dfrac{z_1}{z_2}\right| = \dfrac{1}{2}$ from $-\dfrac{\sqrt{3}}{4} - \dfrac{i}{4}$?

I think the first method is usually easier, all you really have to do is to rewrite the complex numbers from the algebraic to the r,θ form, as you learned in Chapter 4.

4. London. A Level. Core Syllabus.

$z_1 = 1 + i \qquad z_2 = \sqrt{3} - i$

thus $|z_1| = \sqrt{2}$ and $\arg z_1 = \dfrac{\pi}{4}$ (as usual, think or sketch an Argand diagram)

N.B. Use radians for angle measurement, because of the last part of the question.

and $|z_2| = 2$ and $\arg z_2 = -\dfrac{\pi}{6}$

Using $\arg\left(\dfrac{z_1}{z_2}\right) = \arg z_1 - \arg z_2$, to show $\arg\left(\dfrac{z_1}{z_2}\right) = \dfrac{5\pi}{12}$,

then $\arg\left(\dfrac{z_1}{z_2}\right) = \dfrac{\pi}{4} - \left(-\dfrac{\pi}{6}\right) = \dfrac{5\pi}{12}$

6 The four operations on the Argand diagram

If you think of z_1 and z_2 on the Argand diagram as 2-dimensional vectors (which is after all what they are), then the operations become logical and sensible.

1. Addition

We will start with an example, and will choose z_1 and z_2 so that they both fall in the 1st quadrant, which will make your understanding of the principles that much easier. Of course, this works **whatever** quadrants z_1 and z_2 are in.

Let $z_1 = 4 + i$ and $z_2 = 2 + 3i$, so that $z_1 + z_2$ (algebraically) $= 6 + 4i$. On the Argand diagram, z_1 and z_2 are as shown, and $z_1 + z_2$ is the diagonal of the parallelogram OAPB, $\overrightarrow{OP} = z_1 + z_2$, where P is the point (6,4) as expected.

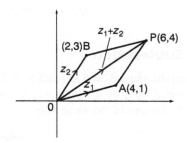

Thus this process is exactly as early vector work which you have already met. If, by any chance, you haven't either met or understood early vector work, this should still make sense to you. It is logical and sensible isn't it?

Of course, this technique will work wherever z_1 and z_2 occur on the Argand diagram, where A and B are. Just locate P by completing the parallelogram OAPB.

We also need to be able to show that

$$|z_1| + |z_2| \geq |z_1 + z_2|$$

where we can use the diagram above, where

$$|z_1| = OA, \qquad |z_2| = OB = AP, \qquad |z_1 + z_2| = OP$$

In any triangle, OAP, the sum of the lengths of two of the sides > the third side, i.e. OA + AP > OP; i.e. $|z_1| + |z_2| > |z_1 + z_2|$.

The only time that OA + AP = OP is in the limiting case when OA, OB and AP are **all** in the same direction, when triangle OAP becomes a straight line, and OA + AP = OP, i.e. $|z_1| + |z_2| = |z_1 + z_2|$: thus

$$|z_1| + |z_2| \geq |z_1 + z_2| \tag{1}$$

N.B. We are only talking about length here (i.e. modulus). Direction (i.e. argument) is irrelevant.

N.B. Also in any triangle OAP, since the sum of the two sides > the third side, it can also be said that OP + PA > OA.

$$|z_1 + z_2| + |z_2| > |z_1|$$

i.e.

$$|z_1 + z_2| > |z_1| - |z_2|$$

and also PO + OA > PA

$$|z_1 + z_2| + |z_1| > |z_2| \Rightarrow |z_1 + z_2| > |z_2| - |z_1|$$

i.e.

$$|z_1 + z_2| > ||z_1| - |z_2||$$

and in limiting conditions $|z_1 + z_2| = ||z_1| - |z_2||$

or

$$|z_1 + z_2| \geq ||z_1| - |z_2|| \tag{2}$$

and by (1) and (2),

$$|z_1| + |z_2| \geq |z_1 + z_2| \geq ||z_1| - |z_2||$$

2. Subtraction

Using the same $z_1 = 4 + i$ and $z_2 = 2 + 3i$, $z_1 - z_2$ can be illustrated in exactly the same way, by drawing $-z_2$ as shown in the diagram, and completing the parallelogram in the same way.

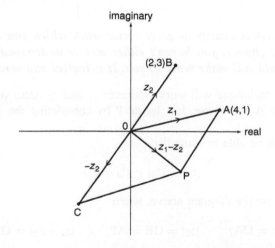

Algebraically, $z_1 - z_2 = 2 - 2i$, and on the diagram on the previous page it can be seen that OP $= z_1 - z_2 = 2 - 2i$ and that P is the point $(2,-2)$.

It is easy! It is not necessary to be frightened of it! I am being a bit pedantic here, because I know that some of you do get frightened!

N.B. Using the same technique as on the previous page, and using the diagram on the previous page, in triangle OAP, where OA (length of OA) $= |z_1|$ AP $= OC = |z_2|$ and OP $= |z_1 - z_2|$, then since OA $+$ AP $\geq$ OP, $|z_1| + |z_2| \geq |z_1 - z_2|$.

Multiplication and division on the Argand diagram are most easily achieved using the results of Chapter 5; i.e.

$$|z_1 z_2| = |z_1||z_2|$$

$$\arg(z_1 z_2) = \arg z_1 + \arg z_2$$

So, on your sketch, you can make sure the modulus of the product equals the **product** of the two individual moduli and the argument of the product equals the **sum** of the two individual arguments.

It will obviously be easier to sketch if in the r, θ form, but this is not necessary.

If $z_1 = 2 + 2i$, $z_2 = 1 + \sqrt{3}i$, as on page 34, then, using the r, θ form, $|z_1| = 2\sqrt{2}$ and $\arg z_1 = \pi/4$ and $|z_2| = 2$, and $\arg z_2 = \pi/3$ (which should be easy for you to do by now; you shouldn't have to look back), then $|z_1 z_2| = 4\sqrt{2}$ and $\arg(z_2 z_2) = 7\pi/12$.

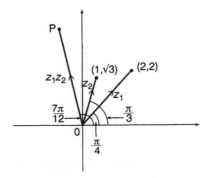

This is hard to show clearly, and so it is not often asked for unless the numbers used are convenient. Sometimes, for example, they ask you to show z, z^2 and z^3 on one diagram, where $z = \sqrt{3} + i$ for instance (see Exercise 6.1, Q1).

Division, at this stage, we will not try to illustrate for the same reasons. You **could** show z_1/z_2, using the above values. I recommend **another** diagram, and obviously the same technique will enable you to show z_1/z_2.

Exercise 6.1

1. If $z = \sqrt{3} + i$, display and label clearly on an Argand diagram, z, z^2 and z^3.

2. If $z_1 = 24 + 7i$ and $|z_2| = 6$, find the greatest and least values of $|z_1 + z_2|$.

3. The two complex numbers z_1 and z_2 are represented on an Argand diagram. Show that $|z_1 + z_2| \leq |z_1| + |z_2|$.

   If $|z_1| = 6$ and $z_2 = 4 + 3i$, show that the greatest value of $|z_1 + z_2|$ is 11 and find its least value.

Exercise 6.1

Worked answers

1. $z = \sqrt{3} + i = 2\left(\dfrac{\sqrt{3}}{2} + \dfrac{1}{2}i\right) = 2\left(\cos\dfrac{\pi}{6} + i\sin\dfrac{\pi}{6}\right)$

   thus $z^2 = z \times z = 4\left(\cos\dfrac{\pi}{3} + i\sin\dfrac{\pi}{3}\right)$ using Chapter 5.

   and $z^3 = z^2 \times z = 8\left(\cos\dfrac{\pi}{2} + i\sin\dfrac{\pi}{2}\right)$ using Chapter 5.

   so that $z = \sqrt{3} + i$ $z^2 = 2 + 2\sqrt{3}i$ $z^3 = 8i$

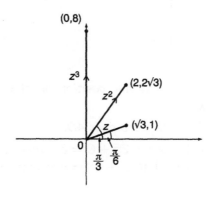

2. $z_1 = 24 + 7i$ thus $|z_1| = 25$ (i.e. $\sqrt{(24^2 + 7^2)}$)

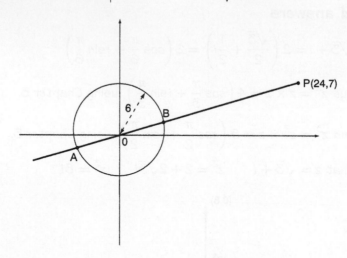

$|z_2| = 6 \Rightarrow z_2$ could be at any position on the circle radius 6 of centre O as shown.

Thus $|z_1 + z_2| \leq |z_1| + |z_2|$ see page 40.

And $|z_1 + z_2| = |z_1| + |z_2| = 25 + 6 = 31$ when z_2 is at position A and the diagram shows that $|z_1 + z_2| = $ distance $AP = 31 = $ greatest value.

And when z_2 is at position B, then least value of $|z_1 + z_2| = BP = |z_1| - |z_2| = 25 - 6 = 19$.

3. See pages 39 and 40 for proof required for first part of question.

Thus as in previous question, $|z_1| = 6$ and $|z_2| = |(4^2 + 3^2) = 5$.

Therefore greatest value of $|z_1 + z_2| = 6 + 5 = 11$

and least value of $|z_1 + z_2| = 6 - 5 = 1$.

7 Useful facts

Useful facts that have occurred within the work of the past pages, but which you may have missed, or not appreciated the value of.

1. $(a + ib)(a - ib) = a^2 + b^2$.
2. If one root of a quadratic equation with real coefficients is complex, then so is the other and these two roots are the complex conjugates of one another.
3. Therefore it follows that for **any** equation with real coefficients, if one root is complex, $a + ib$ say, then an other root will be $a - ib$.
4. $$\frac{1}{\cos \theta + i \sin \theta} = \cos \theta - i \sin \theta: \text{ i.e.}$$

$$\left\{ \frac{1}{\cos \theta + i \sin \theta} = \frac{1(\cos \theta - i \sin \theta)}{(\cos \theta + i \sin \theta)(\cos \theta - i \sin \theta)} = \frac{\cos \theta - i \sin \theta}{\cos^2 \theta + \sin^2 \theta} \right.$$

$$\left. = \frac{\cos \theta - i \sin \theta}{1} \right\}$$

5. Similarly,

$$\frac{1}{\cos \theta - i \sin \theta} = \cos \theta + i \sin \theta$$

6. $\cos \theta - i \sin \theta = \cos(-\theta) + i \sin(-\theta)$.

(**Think about it. Use your trigonometric $\cos \theta = \cos(-\theta)$ and $\sin \theta = -\sin(-\theta)$. If you are still worried, think about a particular angle: $\cos 30$ and $\cos(-30)$ for example.**)

Using 3, try this question.

Exercise 7.1

1. Given that one root of the equation

$$z^4 - 6z^3 + 23z^2 - 34z + 26 = 0 \text{ is } 1 + i, \text{ find the others.}$$

Exercise 7.1

Worked answers

1. $z^4 - 6z^3 + 23z^2 - 34z + 26 = 0$

    One root is $z = 1 + i$ thus another root is $z = 1 - i$

    A quadratic factor is $z^2 - 2z + 2$ (using $x^2 - (\alpha + \beta)x + (\alpha\beta) = 0$)

    $z^4 - 6z^3 + 23z^2 - 34z + 26 = 0$ can be factorized to

    $(z^2 - 2z + 2)(z^2 - 2z + 2)(\qquad\qquad) = 0$

    To locate the second bracket, start by using common sense and algebracic skill:

    $$(z^2 - 2z + 2)(z^2 \ldots\ldots\ldots)$$

    Now start multiplying:

    $$z^4 - 2z^3 + 2z^2 \ldots\ldots\ldots$$

    z^4 is correct. We need $-6z^3$ and we have so far $-2z^3$, so we need $-4z^3$. In the second bracket, put $-4z$, which will multiply with z^2 to give $-4z^3$. We now have

    $$(z^2 - 2z + 2)(z^2 - 4z \ldots\ldots)$$

    Now the number at the end of the second bracket must be $+13$ to give $+26$. Now multiply the two brackets together and they should give the original quartic

    i.e. $(z^2 + 2z + 2)(z^2 - 4z + 13) = 0$

    and $z^4 - 2z^3 - 4z^3 + 13z^2 + 2z^2 + 8z^2 - 26z - 8z + 26 = 0$

    so $z^4 - 6z^3 + 23z^2 + 26 = 0$, which is the required quartic.

    This second quadratic gives roots

    $$z = \frac{4 \pm \sqrt{16 - 52}}{2} = \frac{4 \pm \sqrt{-36}}{2} = 2 \pm 3i$$

    Therefore the other roots of the quadratic are $z = 1 - i$, $2 + 3i$, $2 - 3i$.

8 A sample question

London.

14. Find the modulus and argument of each of the complex numbers z_1 and z_2, where

$$z_1 = \frac{1+i}{1-i}, \ z_2 = \frac{\sqrt{2}}{1-i}$$

Plot the points representing z_1, z_2 and $z_1 + z_2$ on an Argand diagram. Deduce from your diagram that

$$\tan(3\pi/8) = 1 + \sqrt{2}$$

$$z_1 = \frac{1+i}{1-i} = \frac{(1+i)(1+i)}{(1-i)(1+i)} = \frac{1+2i+i^2}{1+1} = \frac{1+2i-1}{2} = i$$

thus $|z_1| = 1$ and $\arg z_1 = \pi/2$

$$z_2 = \frac{\sqrt{2}}{1-i} = \frac{\sqrt{2}(1+i)}{(1-i)(1+i)} = \frac{\sqrt{2}}{2}(1+i) = \frac{1}{\sqrt{2}} + \frac{1}{\sqrt{2}}i$$

thus $|z_2| = 1$ and $\arg z_2 = \pi/4$.

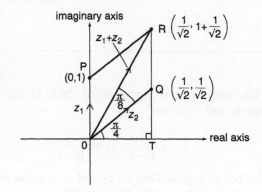

From the diagram, using triangle OTR,

$$\tan \angle TOR = \frac{1 + \dfrac{1}{\sqrt{2}}}{\dfrac{1}{\sqrt{2}}} = \sqrt{2} + 1$$

i.e.

$$\tan \frac{3\pi}{8} = 1 + \sqrt{2}$$

Now you try the question in Chapter 9.

9 Further questions

Q1. London.

Given that the real and imaginary parts of the complex number $z = x + iy$ satisfy the equation:

$$(2 - i)x - (1 + 3i)y - 7 = 0, \text{ find } x \text{ and } y$$

State the values of (a) $|z|$, (b) $\arg z$.

Q2. London.

Given that $z = 2 - i$, show that $z^2 = 3 - 4i$.
Hence, or otherwise, find the roots z_1 and z_2, of the equation

$$(z + i)^2 = 3 - 4i$$

Display these roots on an Argand diagram.
(a) Deduce that $|z_1 - z_2| = 2\sqrt{5}$.
(b) Find the value of $\arg(z_1 + z_2)$.

Q3. London.

Given that $z = \dfrac{1}{2 + i}$ express in the form $a + bi$, where $a, b \in R$,

(a) z^2,　　(b) $z - \dfrac{1}{z}$.

Find the modulus of z^2.
Determine the argument of $(z - (1/z))$, giving your answer in degrees to one decimal place.

Q4. AEB.

Find, in terms of π, the argument of the complex numbers
(a) $(1 + i)^2$

(b) $\dfrac{3 + i}{1 + 2i}$

(c) $\left(\cos \dfrac{\pi}{3} + i \sin \dfrac{\pi}{3} \right) \left(\cos \dfrac{\pi}{4} - i \sin \dfrac{\pi}{4} \right)$

Q5. AEB.

Given that $z = a + ib$, where a and b are real, and that z satisfies the equation:

$$(2 + i)(z + 3i) = 7i - 6$$

find the values of a and b.

Show z on an Argand diagram, stating the modulus and the argument of z.

Q6. AEB.

Find the values of the real numbers a and b so that

$$(a + ib)^2 = 16 - 30i$$

Write down the square roots of $16 - 30i$.

Q7. AEB.

Given that $z = 2 + 2i$, express z in the form $r(\cos\theta + i\sin\theta)$, where r is a positive real number and $-\pi < \theta \leq \pi$.

On the same Argand diagram, display and label clearly the numbers

$$z, \ z^2 \text{ and } \frac{4}{z}$$

Find the values of $|z + z^2|$ and $\arg\left(z + \dfrac{4}{z}\right)$.

Worked answers

Q1. London.

$z = x + iy$

$(2 - i)x - (1 + 3i)y - 7 = 0$

To find x and y:

$2x - ix - y - 3iy - 7 = 0$

Equating real parts $2x - y - 7 = 0$ (1)

Equating imaginary parts $-x - 3y = 0$ thus $x = -3y$ (2)

and substituting into (1) $-6y - y - 7 = 0 \Rightarrow y = -1 \Rightarrow x = 3$

Thus $z = 3 - i$

(a) $|z| = \sqrt{(3^2 + (-1)^2)} = \sqrt{10}$

(b) $\arg z = \arctan\left(-\dfrac{1}{3}\right)$ or $-\arctan\dfrac{1}{3}$ (4th quadrant)

      $= -18.4°$

Q2. London.

Given $z = 2 - i$, then $z^2 = (2 - i)(2 - i) = 4 - 4i + i^2 = 3 - 4i$

if $(z + i)^2 = 3 - 4i$, then $z + i = \pm(2 - i)$

i.e. $z = 2 - i - i$ or $z = -(2 - i) - i$

i.e. $z_1 = 2 - 2i$ or $z_2 = -2$

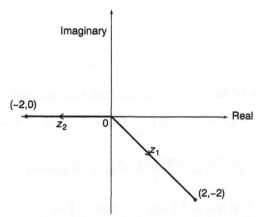

(a) $z_1 - z_2 = (2 - 2i) - (-2)$

           $= 4 - 2i$

$$|z_1 - z_2| = \sqrt{(16 + 4)} = \sqrt{20} = \sqrt{(4 \times 5)}$$

$$|z_1 - z_2| = 2\sqrt{5}$$

(b) $z_1 + z_2 = (2 - 2i) + (-2) = -2i$

thus $\arg(z_1 + z_2) = -\dfrac{\pi}{2}$

Q3. London.

$$z = \frac{1}{2 + i} = \frac{1}{(2 + i)} \frac{(2 - i)}{(2 - i)} = \frac{2 - i}{5} = \frac{2}{5} - \frac{1}{5}i$$

(a) $z^2 = \dfrac{1}{25}(2 - i)(2 - i) = \dfrac{1}{25}(4 - 4i + i^2) = \dfrac{3}{25} - \dfrac{4i}{25}$

(b) $z - \dfrac{1}{z} = \dfrac{2}{5} - \dfrac{1}{5}i - (2 + i)$

$$= -\frac{8}{5} - \frac{6}{5}i$$

$$|z^2| = \left|\frac{3}{25} - \frac{4}{25}i\right| = \frac{1}{25}|3 - 4i| = \frac{1}{25}\sqrt{(3^2 + 4^2)} = \frac{5}{25} = \frac{1}{5}$$

$$\arg\left(z - \frac{1}{z}\right) = \arg\left(-\frac{8}{5} - \frac{6}{5}i\right)$$

$$= -180° + \arctan\frac{6}{8}$$

$$= -180° + 36.9° = -143.1° \quad (\text{3rd quadrant})$$
$$\text{to 1 decimal place}$$

Q4. AEB.

To find, in terms of π, arguments of

(a) $(1 + i)^2 = (1 + i)(1 + i) = 1 - 1 + 2i = 2i.$

$\arg(1 + i)^2 = \dfrac{\pi}{2}$

(b) $\dfrac{3 + i}{1 + 2i} = \dfrac{(3 + i)(1 - 2i)}{(1 + 2i)(1 - 2i)} = \dfrac{3 + 2 + i - 6i}{1 + 4} = \dfrac{5 - 5i}{5} = 1 - i$

thus $\arg(1 - i) = -\dfrac{\pi}{4}$ (use Argand diagram)

(c) $\left(\cos\dfrac{\pi}{3} + i\sin\dfrac{\pi}{3}\right)\left(\cos\dfrac{\pi}{4} - i\sin\dfrac{\pi}{4}\right)$

$$= \left(\cos\frac{\pi}{3} + i\sin\frac{\pi}{3}\right)\left(\cos\left(-\frac{\pi}{4}\right) + i\sin\left(-\frac{\pi}{4}\right)\right)$$

$$= \cos\left(\frac{\pi}{3} - \frac{\pi}{4}\right) + i\sin\left(\frac{\pi}{3} - \frac{\pi}{4}\right)$$

$$= \cos\frac{\pi}{12} + i\sin\frac{\pi}{12}$$

thus $\arg\left(\cos\dfrac{\pi}{12} + i\sin\dfrac{\pi}{12}\right) = \dfrac{\pi}{12}$

Q5. AEB.

$(2 + i)(z + 3i) = 7i - 6$ where $z = a + ib$

to find a and b.

$$z + 3i = \frac{-6 + 7i}{2 + i} = \frac{(-6 + 7i)(2 - i)}{(2 + i)(2 - i)} = \frac{-12 + 7 + 14i + 6i}{4 + 1}$$

$$= \frac{-5 + 20i}{5} = -1 + 4i$$

thus $z + 3i = -1 + 4i$

so that $z = -1 + i = a + ib$ so that $a = -1$, $b = 1$

$$|z| = \sqrt{2}$$

$$\arg z = \frac{3\pi}{4}$$

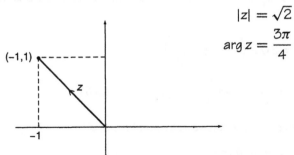

Q6. AEB.

$(a + ib)^2 = 16 - 30i$, a and b real.

$a^2 - b^2 + 2abi = 16 - 30i$

$a^2 - b^2 = 16$ (1) $2ab = -30$ (2)

From (2), $ab = -15 \Rightarrow a = -\dfrac{15}{b}$

substituting in (1) $\left(-\dfrac{15}{b}\right)^2 - b^2 = 16 \Rightarrow \dfrac{225}{b^2} - b^2 = 16$

$$\Rightarrow b^4 + 16b^2 - 225 = 0$$

$$(b^2 + 25)(b^2 - 9) = 0$$

$b^2 = 9$ ($b^2 \neq -25$, because b is real)

$\quad b = 3, -3$

thus $a = -5, 5$.

when $b = 3$, $a = -5$ and when $b = -3$, $a = 5$

thus the two square roots of $16 - 30i$ are $(-5 + 3i)$ and $(5 - 3i)$

Q7. $z = 2 + 2i = 2\sqrt{2}\left(\dfrac{1}{\sqrt{2}} + \dfrac{1}{\sqrt{2}}i\right)$

$z = 2\sqrt{2}\left(\cos\dfrac{\pi}{4} + i\sin\dfrac{\pi}{4}\right) = r(\cos\theta + i\sin\theta)$

$r = 2\sqrt{2}$ which is a positive real number, and

$\theta = \dfrac{\pi}{4}$ and $-\pi < \theta \leq \pi$.

$z^2 = z \cdot z = 2\sqrt{2}\left(\cos\dfrac{\pi}{4} + i\sin\dfrac{\pi}{4}\right) 2\sqrt{2}\left(\cos\dfrac{\pi}{4} + i\sin\dfrac{\pi}{4}\right)$

$\qquad = 8\left(\cos\dfrac{\pi}{2} + i\sin\dfrac{\pi}{2}\right) = 8i$

$\dfrac{4}{z} = \dfrac{4}{2 + 2i} = \dfrac{2}{1 + i} = \dfrac{2(1 - i)}{(1 + i)(1 - i)} = \dfrac{2(1 - i)}{2} = 1 - i$

$z = 2 + 2i$, point A

$z^2 = 8i$, point B

$\dfrac{4}{z} = 1 - i$, point C

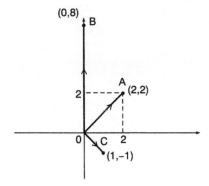

$z + z^2 = (2 + 2i) + 8i = 2 + 10i,$

thus $|z + z^2| = \sqrt{(2^2 + 10^2)} = \sqrt{104} = 2\sqrt{26}$

$z + \dfrac{4}{z} = (2 + 2i) + (1 - i) = 3 + i,$

thus $\arg\left(z + \dfrac{4}{z}\right) = \arctan\dfrac{1}{3} = 18.43°$

10 De Moivre's theorem

This theorem states that, for any rational value of n, one value of $(\cos\theta + i\sin\theta)$ is given by:

$$(\cos\theta + i\sin\theta)^n = \cos n\theta + i\sin n\theta$$

N.B. The reason for saying 'one value' is that there is more then one value for expressions like $(\cos\theta + i\sin\theta)^{2/3}$, as I will show you later (see Chapters 13, 14 and 15).

Here follow proofs of De Moivre's theorem:

(a) when n is a positive integer;
(b) when n is a negative integer; and
(c) when n is rational.

Proof of this theorem (a) when n is a positive integer.
This is done by induction, is quite easy and is quite often asked for in A Level papers, and is like this:

(i) Assume true when $n = k$; i.e.

$$(\cos\theta + i\sin\theta)^k = \cos k\theta + i\sin k\theta \qquad (1)$$

(ii) Multiply both sides by $\cos\theta + i\sin\theta$; i.e.

$$(\cos\theta + i\sin\theta)^k(\cos\theta + i\sin\theta) = (\cos k\theta + i\sin k\theta)(\cos\theta + i\sin\theta)$$

and work out the RHS as follows:

$$(\cos\theta + i\sin\theta)^{k+1} = (\cos k\theta\cos\theta - \sin k\theta\sin\theta)$$
$$+ i(\sin k\theta\cos\theta + \cos k\theta\sin\theta)$$
$$= \cos(k+1)\theta + i\sin(k+1)\theta \qquad (2)$$

(iii) Therefore *if* (1) is true, then so is (2), as both are in the same form.

(iv) But $(\cos\theta + i\sin\theta)^2 = (\cos\theta + i\sin\theta)(\cos\theta + i\sin\theta)$
$$= (\cos^2\theta - \sin^2\theta) + 2i\sin\theta\cos\theta$$
$$= \cos 2\theta + i\sin 2\theta$$

Thus when $n = 2$, $(\cos\theta + i\sin\theta)^2 = \cos 2\theta + i\sin 2\theta$

(v) Using (iii), if $n = 2$ is true, then so is $n = 3$, and if $n = 3$ is true, then so is $n = 4$ and so on.

(vi) The theorem is true for all positive integral values of n; i.e.

$$(\cos\theta + i\sin\theta)^7 = \cos 7\theta + i\sin 7\theta$$

This is a very significant result—just think of $(a + b)^7$, this has 8 terms!

Proof of this theorem (b) when n is a negative integer.

(i) Consider $n = -m$ where m is a positive integer:

$$(\cos\theta + i\sin\theta)^n = (\cos\theta + i\sin\theta)^{-m}$$

$$= \frac{1}{(\cos\theta + i\sin\theta)^m}$$

$$= \frac{1}{(\cos m\theta + i\sin m\theta)} \quad \text{(using the result of theorem (a))}$$

(ii) Use the complex conjugate on the (top and bottom) numerator and denominator, as follows:

$$\frac{1}{(\cos m\theta + i\sin m\theta)} = \frac{1(\cos m\theta - i\sin m\theta)}{(\cos m\theta + i\sin m\theta)(\cos m\theta - i\sin m\theta)}$$

$$= \frac{\cos m\theta - i\sin m\theta}{\cos^2 m\theta + \sin^2 m\theta}$$

$$= \frac{\cos m\theta - i\sin m\theta}{1}$$

$$= \cos(-m\theta) + i\sin(-m\theta)$$

(N.B. See page 45, Chapter 7, No. 6, if you need more advice here.)

(iii) From (i) and (ii), $\cos(-m\theta) + i\sin(-m\theta) = \cos n\theta + i\sin n\theta$
therefore $(\cos\theta + i\sin\theta)^n = \cos n\theta + i\sin n\theta$ when n is a negative integer.

Proof of this theorem (c) when n is rational: i.e. $n = p/q$, where p and q are integers.

i.e. to prove

$$(\cos\theta + i\sin\theta)^{p/q} = \cos\frac{p\theta}{q} + i\sin\frac{p\theta}{q}$$

(i) $(\text{RHS})^q = (\cos(p\theta/q) + i\sin(p\theta/q))^q = \cos p\theta + i\sin p\theta$ (as q is an integer and already proved in (a) or (b)).

(ii) and $(\cos p\theta + i\sin p\theta) = (\cos\theta + i\sin\theta)^p$ (as p is an integer, thus again using (a) or (b)).

(iii) $\left(\cos\dfrac{p\theta}{q} + i\sin\dfrac{p\theta}{q}\right)^q = (\cos\theta + i\sin\theta)^p$

$$\Rightarrow \left(\cos\frac{p\theta}{q} + i\sin\frac{p\theta}{q}\right) = (\cos\theta + i\sin\theta)^{p/q}$$

Therefore De Moivre's theorem is proved for any rational value of n.

N.B. (1) $\cos n\theta + i \sin n\theta$ *is only* **one** *value of* $(\cos\theta + i\sin\theta)^n$ *when* $n = p/q$. *There are other values as you will see later (i.e.* $z^5 = 1$ *has 5 values for* z. *Only one of them is* $z = 1$. *There are four others. De Moivre's theorem will enable us to find them (see Chapters 13, 14 and 15).)*

N.B. (2) If you are asked to prove this theorem, it will probably be for (a) n *a positive integer only. This is really an opportunity for the examiners to check your knowledge and handling of the proof by induction. Make sure you can do this proof (see pages 57 and 58).*

N.B. (3) You have proved De Moivre's theorem,

$$(\cos\theta + i\sin\theta)^n = \cos n\theta + i\sin n\theta, \text{ and } \textbf{nothing } \text{else.}$$

If $\cos\theta - i\sin\theta$ *appears, you* **must** *rewrite it as* $\cos(-\theta) + i\sin(-\theta)$ *and* **then** *you can use De Moivre's theorem.*

Uses of De Moivre's theorem are many, and will follow on the next pages. First try the following exercise, using the De Moivre result which you can assume.

Exercise 10.1

Simplify, using the De Moivre's theorem, and giving the answers in the $a + ib$ form.

1. $(\cos\theta + i\sin\theta)^7$

2. $\dfrac{1}{(\cos\theta + i\sin\theta)^3}$

3. $(\cos\theta - i\sin\theta)^{-3}$

4. $(\cos\theta + i\sin\theta)^3(\cos\theta + i\sin\theta)^2$

5. $\left(\cos\dfrac{\pi}{6} + i\sin\dfrac{\pi}{6}\right)^3$

6. $\left(\cos\dfrac{\pi}{4} + i\sin\dfrac{\pi}{4}\right)^{-2}$

7. $(\cos 3\theta + i\sin 3\theta)(\cos 5\theta + i\sin 5\theta)$

8. $\dfrac{(\cos 9\theta + i\sin 9\theta)}{(\cos 2\theta - i\sin 2\theta)}$

9. $\dfrac{(\cos\theta + i\sin\theta)}{(\cos 2\theta - i\sin 2\theta)^2}$

10. $\dfrac{(\cos 2\theta + i\sin 2\theta)^3}{(\cos\theta + i\sin\theta)^2}$

11. Simplify, without the use of a calculator,

$$\frac{\left(\cos\dfrac{1}{7}\pi - i\sin\dfrac{1}{7}\pi\right)^3}{\left(\cos\dfrac{1}{7}\pi + i\sin\dfrac{1}{7}\pi\right)^4}$$

This is a past A Level question from a University of London paper.

Exercise 10.1

Worked answers

The following questions are done using

(a) De Moivre's theorem for n rational:

$$(\cos\theta + i\sin\theta)^n = \cos n\theta + i\sin n\theta$$

(b) $\dfrac{1}{\cos\theta + i\sin\theta} = \cos\theta - i\sin\theta$ and therefore

$$\dfrac{1}{\cos\theta - i\sin\theta} = \cos\theta + i\sin\theta$$

(c) $\cos(-\theta) = \cos\theta$ and $\sin(-\theta) = -\sin\theta \Rightarrow \cos\theta - i\sin\theta$

$$= \cos(-\theta) + i\sin(-\theta)$$

There are, of course, several ways to do these. For example in Q2 I have used two different ways. If ever you prefer to do a question another way, you can of course do so.

Most of these questions **can** be done with no working, especially as you get more practice at doing them. I recommend that you **do** write down a little. No working and a **wrong** answer will earn you **no** marks!

I will make **one** deliberate mistake. Can you find it? I will dislike doing it, and will tell you which it is at the end of the exercise!

1. $(\cos\theta + i\sin\theta)^7 = \cos 7\theta + i\sin 7\theta$ by De Moivre's theorem

2. $\dfrac{1}{(\cos\theta + i\sin\theta)^3} = \dfrac{1}{\cos 3\theta + i\sin 3\theta} = \cos 3\theta - i\sin 3\theta$

   OR

   $\dfrac{1}{(\cos\theta + i\sin\theta)^3} = (\cos\theta + i\sin\theta)^{-3} = \cos(-3\theta) + i\sin(-3\theta)$

   $$= \cos 3\theta - i\sin 3\theta$$

3. $(\cos\theta - i\sin\theta)^{-3} = (\cos(-\theta) + i\sin(-\theta))^{-3}$

   $$= \cos(-3)(-\theta) + i\sin(-3)(-\theta)$$

   $$= \cos 3\theta + i\sin 3\theta$$

4. $(\cos\theta + i\sin\theta)^3(\cos\theta + i\sin\theta)^2 = (\cos\theta + i\sin\theta)^6$

   $$= \cos 6\theta + i\sin 6\theta$$

5. $\left(\cos\dfrac{\pi}{6} + i\sin\dfrac{\pi}{6}\right)^3 = \cos\dfrac{3\pi}{6} + i\sin\dfrac{3\pi}{6} = \cos\dfrac{\pi}{2} + i\sin\dfrac{\pi}{2} = i$

6. $\left(\cos\dfrac{\pi}{4} + i\sin\dfrac{\pi}{4}\right)^{-2} = \cos\left(-\dfrac{2\pi}{4}\right) + i\sin\left(-\dfrac{2\pi}{4}\right)$

$$= \cos\left(-\dfrac{\pi}{2}\right) + i\sin\left(-\dfrac{\pi}{2}\right) = -i$$

7. $(\cos 3\theta + i\sin 3\theta)(\cos 5\theta + i\sin 5\theta) = \cos(3\theta + 5\theta) + i\sin(3\theta + 5\theta)$

$$= \cos 8\theta + i\sin 8\theta$$

(remembering that, when two complex numbers in the r,θ form are multiplied, their arguments are added)

OR

$$(\cos 3\theta + i\sin 3\theta)(\cos 5\theta + i\sin 5\theta)$$

$$= (\cos\theta + i\sin\theta)^3(\cos\theta + i\sin\theta)^5$$

$$= (\cos\theta + i\sin\theta)^8 = \cos 8\theta + i\sin 8\theta$$

8. $\dfrac{\cos 9\theta + i\sin 9\theta}{\cos 2\theta - i\sin 2\theta} = (\cos 9\theta + i\sin 9\theta)(\cos 2\theta + i\sin 2\theta)$

$$= \cos 11\theta + i\sin 11\theta$$

9. $\dfrac{\cos\theta + i\sin\theta}{(\cos 2\theta - i\sin 2\theta)^2} = (\cos\theta + i\sin\theta)(\cos 2\theta + i\sin 2\theta)^2$

$$= (\cos\theta + i\sin\theta)(\cos 4\theta + i\sin 4\theta)$$

$$= \cos 5\theta + i\sin 5\theta$$

10. $\dfrac{(\cos 2\theta + i\sin 2\theta)^3}{(\cos\theta + i\sin\theta)^2} = \dfrac{(\cos\theta + i\sin\theta)^6}{(\cos\theta + i\sin\theta)^2}$

$$= (\cos\theta + i\sin\theta)^4 = \cos 4\theta + i\sin 4\theta$$

11. $\dfrac{\left(\cos\dfrac{1}{7}\pi - i\sin\dfrac{1}{7}\pi\right)^3}{\left(\cos\dfrac{1}{7}\pi + i\sin\dfrac{1}{7}\pi\right)^4} = \dfrac{1}{\left(\cos\dfrac{1}{7}\pi + i\sin\dfrac{1}{7}\pi\right)^4\left(\cos\dfrac{1}{7}\pi + i\sin\dfrac{1}{7}\pi\right)^3}$

$$= \dfrac{1}{\left(\cos\dfrac{1}{7}\pi + i\sin\dfrac{1}{7}\pi\right)^7}$$

$$= \dfrac{1}{\cos\pi + i\sin\pi} = \dfrac{1}{-1} = -1$$

The deliberate mistake is in Q4, and is a **very** common error.

$$(\cos\theta + i\sin\theta)^3(\cos\theta + i\sin\theta)^2 = (\cos\theta + i\sin\theta)^5$$

$$= \cos 5\theta + i\sin 5\theta$$

11 Use of De Moivre's theorem I

In this chapter we will use De Moivre's theorem to prove trigonometric formulae where $\sin n\theta$ (n positive integer) is expressed in terms of powers of $\sin\theta$ only, and similarly for $\cos n\theta$ and $\tan n\theta$, e.g.

$$\cos 5\theta \equiv 16\cos^5\theta - 20\cos^3\theta + 5\cos\theta$$

First, I will do an example for you:
To prove that

$$\tan 3\theta \equiv \frac{3\tan\theta - \tan^3\theta}{1 - 3\tan^2\theta}$$

Because this is only $\tan 3\theta$, and as you should know suitable trigonometric formulae, you could prove this without the help of De Moivre's theorem. However, for higher multiple angles, as in Exercise 11.1 on the next page, the following method is the easiest and therefore a sensible way to proceed.

$$(\cos\theta + i\sin\theta)^3 \equiv \cos 3\theta + i\sin 3\theta \qquad \text{(De Moivre's theorem) (1)}$$

$$(\cos\theta + i\sin\theta)^3 \equiv \cos^3\theta + 3\cos^2\theta(i\sin\theta) + 3\cos\theta(i^2\sin^2\theta) + i^3\sin^3\theta$$

$$\equiv \cos^3\theta - 3\cos\theta\sin^2\theta + i(3\cos^2\theta\sin\theta - \sin^3\theta) \qquad (2)$$

(Binomial theorem or Pascal's triangle)

Equating real and imaginary parts of (1) and (2)

$$\cos 3\theta \equiv \cos^3\theta - 3\cos\theta\sin^2\theta \qquad (3)$$

and $\sin 3\theta \equiv 3\cos^2\theta\sin\theta - \sin^3\theta \qquad (4)$

Thus $\dfrac{(4)}{(3)} = \dfrac{\sin 3\theta}{\cos 3\theta} \equiv \dfrac{3\cos^2\theta\sin\theta - \sin^3\theta}{\cos^3\theta - 3\cos\theta\sin^2\theta},$

and dividing top and bottom by $\cos^3\theta$,

$$\tan 3\theta \equiv \frac{\dfrac{3\cos^2\theta\sin\theta}{\cos^3\theta} - \dfrac{\sin^3\theta}{\cos^3\theta}}{\dfrac{\cos^3\theta}{\cos^3\theta} - \dfrac{3\cos\theta\sin^2\theta}{\cos^3\theta}} = \frac{3\dfrac{\sin\theta}{\cos\theta} - \dfrac{\sin^3\theta}{\cos^3\theta}}{1 - 3\dfrac{\sin^2\theta}{\cos^2\theta}}$$

so that $\tan 3\theta \equiv \dfrac{3\tan\theta - \tan^3\theta}{1 - 3\tan^2\theta}$ as required.

N.B. (3) could be written entirely in terms of powers of $\cos\theta$, *and similarly for (4) if required, by using* $\cos^2\theta + \sin^2\theta = 1$.

Now try Exercise 11.1

Exercise 11.1

Prove using De Moivre's theorem that:

1. $\cos 4\theta \equiv 8\cos^4\theta - 8\cos^2\theta + 1$

2. $\sin 5\theta \equiv 5\sin\theta - 20\sin^3\theta + 16\sin^5\theta$

3. $\tan 4\theta \equiv \dfrac{4\tan\theta - 4\tan^3\theta}{1 - 6\tan^2\theta + \tan^4\theta}$

Exercise 11.1

Worked answers

1. To prove that $\cos 4\theta \equiv 8\cos^4\theta - 8\cos^2\theta + 1$

$$(\cos\theta + i\sin\theta)^4 = \cos 4\theta + i\sin 4\theta \qquad \text{De Moivre's theorem (1)}$$

$$(\cos\theta + i\sin\theta)^4 \equiv \cos^4\theta + 4\cos^3\theta(i\sin\theta) + 6\cos^2\theta(i^2\sin^2\theta)$$
$$+ 4\cos\theta(i^3\sin^3\theta) + i^4\sin^4\theta$$
$$\equiv \cos^4\theta + 4i\cos^3\theta\sin\theta - 6\cos^2\theta\sin^2\theta$$
$$- 4i\cos\theta\sin^3\theta + \sin^4\theta \qquad (2)$$

Binomial theorem or Pascal's triangle

Equating the real parts of (1) and (2),

$$\cos 4\theta \equiv \cos^4\theta - 6\cos^2\theta\sin^2\theta + \sin^4\theta$$
$$\equiv \cos^4\theta - 6\cos^2\theta(1 - \cos^2\theta) + (1 - \cos^2\theta)^2$$

using $\cos^2\theta + \sin^2\theta = 1$

$$\equiv \cos^4\theta - 6\cos^2\theta + 6\cos^4\theta + 1 - 2\cos^2\theta + \cos^4\theta$$
$$\cos 4\theta \equiv 8\cos^4\theta - 8\cos^2\theta + 1$$

2. To prove that $\sin 5\theta \equiv 5\sin\theta - 20\sin^3\theta + 16\sin^5\theta$

$$(\cos\theta + i\sin\theta)^5 \equiv \cos 5\theta + i\sin 5\theta \qquad \text{De Moivre's theorem (1)}$$

$$(\cos\theta + i\sin\theta)^5 \equiv \cos^5\theta + 5\cos^4\theta(i\sin\theta) + 10\cos^3\theta(i^2\sin^2\theta)$$
$$+ 10\cos^2\theta(i^3\sin^3\theta) + 5\cos\theta(i^4\sin^4\theta)$$
$$+ i^5\sin^5\theta$$
$$\equiv \cos^5\theta + 5i\cos^4\theta\sin\theta - 10\cos^3\theta\sin^2\theta$$
$$- 10i\cos^2\theta\sin^3\theta + 5\cos\theta\sin^4\theta + i\sin^5\theta \qquad (2)$$

Binomial theorem or Pascal's triangle

Equating the imaginary parts of (1) and (2),

$$\sin 5\theta \equiv 5\cos^4\theta\sin\theta - 10\cos^2\theta\sin^3\theta + \sin^5\theta$$
$$\equiv 5(1 - \sin^2\theta)^2\sin\theta - 10(1 - \sin^2\theta)\sin^3\theta + \sin^5\theta$$

using $\cos^2\theta + \sin^2\theta = 1$

$$\equiv 5(1 - 2\sin^2\theta + \sin^4\theta)\sin\theta - 10\sin^3\theta + 10\sin^5\theta + \sin^5\theta$$
$$\equiv 5\sin\theta - 10\sin^3\theta + 5\sin^5\theta - 10\sin^3\theta + 10\sin^5\theta + \sin^5\theta$$
$$\sin 5\theta \equiv 5\sin\theta - 20\sin^3\theta + 16\sin^5\theta$$

3. To prove that $\tan 4\theta \equiv \dfrac{4\tan\theta - 4\tan^3\theta}{1 - 6\tan^2\theta + \tan^4\theta}$

Look again at Q1, parts (1) and (2)

Equating real and imaginary parts of (1) and (2)

$$\cos 4\theta \equiv \cos^4\theta - 6\cos^2\theta\sin^2\theta + \sin^4\theta$$

$$\text{and } \sin 4\theta \equiv 4\cos^3\theta\sin\theta - 4\cos\theta\sin^3\theta$$

Thus $\dfrac{\sin 4\theta}{\cos 4\theta} \equiv \dfrac{4\cos^3\theta\sin\theta - 4\cos\theta\sin^3\theta}{\cos^4\theta - 6\cos^2\theta\sin^2\theta + \sin^4\theta}$

and dividing top and bottom of the RHS by $\cos^4\theta$,

$$\tan 4\theta \equiv \dfrac{\dfrac{4\cos^3\theta\sin\theta}{\cos^4\theta} - \dfrac{4\cos\theta\sin^3\theta}{\cos^4\theta}}{\dfrac{\cos^4\theta}{\cos^4\theta} - \dfrac{6\cos^2\theta\sin^2\theta}{\cos^4\theta} + \dfrac{\sin^4\theta}{\cos^4\theta}}$$

$$\equiv \dfrac{4\tan\theta - 4\tan^3\theta}{1 - 6\tan^2\theta + \tan^4\theta}$$

12 Use of De Moivre's theorem II

In this chapter we will use De Moivre's theorem to prove trigonometric formulae where $\cos^n \theta$ or $\sin^n \theta$ is expressed in terms of the cosines or sines of multiples of θ; e.g.

$$\cos^4 \theta \equiv \frac{1}{8}(\cos 4\theta + 4 \cos 2\theta + 3)$$

To do this sort of question, the simplest way is as follows:
It has already been shown that, if $z = \cos\theta + i \sin\theta$, then

$$\frac{1}{z} = \cos\theta - i \sin\theta$$

and therefore,

$$z + \frac{1}{z} = 2\cos\theta \tag{1}$$

and

$$z - \frac{1}{z} = 2i \sin\theta \tag{2}$$

also $z^n = (\cos\theta + i \sin\theta)^n = \cos n\theta + i \sin n\theta$

and $\dfrac{1}{z^n} = z^{-n} = (\cos\theta + i \sin\theta)^{-n} = \cos(-n\theta) + i \sin(-n\theta)$

$$= \cos n\theta - i \sin n\theta$$

and therefore

$$z^n + \frac{1}{z^n} = 2\cos n\theta \tag{3}$$

and

$$z^n - \frac{1}{z^n} = 2i \sin n\theta \tag{4}$$

In the example on the previous page:

to prove $\cos^4\theta \equiv \dfrac{1}{8}(\cos 4\theta + 4\cos 2\theta + 3)$:

using (1), $\left(z + \dfrac{1}{z}\right)^4 = (2\cos\theta)^4 = 16\cos^4\theta$

and also, $\left(z + \dfrac{1}{z}\right)^4 = z^4 + 4z^3 \cdot \dfrac{1}{z} + 6z^2 \cdot \dfrac{1}{z^2} + 4z \cdot \dfrac{1}{z^3} + \dfrac{1}{z^4}$

$$= \left(z^4 + \dfrac{1}{z^4}\right) + 4\left(z^2 + \dfrac{1}{z^2}\right) + 6$$

and using (3) $= 2\cos 4\theta + 8\cos 2\theta + 6$

thus $16\cos^4\theta = 2\cos 4\theta + 8\cos 2\theta + 6$

and $\cos^4\theta = \dfrac{1}{8}(\cos 4\theta + 4\cos 2\theta + 3)$ as required.

Now try Exercise 12.1.

Exercise 12.1

1. Prove that $\cos^5\theta = \dfrac{1}{16}(\cos 5\theta + 5\cos 3\theta + 10\cos\theta)$

2. Prove that $\sin^5\theta = \dfrac{1}{16}(\sin 5\theta - 5\sin 3\theta + 10\sin\theta)$

   Hence find $\displaystyle\int (10\sin\theta - 16\sin^5\theta)\,d\theta$

3. Prove $\cos^4\theta + \sin^4\theta = \dfrac{1}{4}(\cos 4\theta + 3)$

4. Find, using De Moivre's theorem $\displaystyle\int_0^{\pi/4} 8\cos^4\theta\,d\theta$

---

*Look again at Exercise 11.1. It is **very important that you realize that these two exercises contain questions that are proved in completely different ways**, as I have shown. If you try to prove questions from either, using the other method, the work involved becomes much longer and more tedious. So, for your own sake, you **must be able to identify these two types of question, and remember how to tackle each.***

Exercise 12.1

Worked answers

1. To prove $\cos^5\theta = \dfrac{1}{16}(\cos 5\theta + 5\cos 3\theta + 10\cos\theta)$

   $z = \cos\theta + i\sin\theta$ thus $z + \dfrac{1}{z} = 2\cos\theta$ and $z^n + \dfrac{1}{z^n} = 2\cos n\theta$

   $$\left(z + \frac{1}{z}\right)^5 = 32\cos^5\theta$$

   $$= z^5 + 5z^4\frac{1}{z} + 10z^3\frac{1}{z^2} + 10z^2\frac{1}{z^3} + 5z\frac{1}{z^4} + \frac{1}{z^5}$$

   $$32\cos^5\theta = \left(z^5 + \frac{1}{z^5}\right) + 5\left(z^3 + \frac{1}{z^3}\right) + 10\left(z + \frac{1}{z}\right)$$

   $$32\cos^5\theta = 2\cos 5\theta + 10\cos 3\theta + 20\cos\theta$$

   $$\cos^5\theta = \frac{1}{16}(\cos 5\theta + 5\cos 3\theta + 10\cos\theta)$$

2. To prove $\sin^5\theta = \dfrac{1}{16}(\sin 5\theta - 5\sin 3\theta + 10\sin\theta)$

   $$z - \frac{1}{z} = 2i\sin\theta \text{ and } z^n - \frac{1}{z^n} = 2i\sin n\theta$$

   $$\left(z - \frac{1}{z}\right)^5 = 32i^5\sin^5\theta = z^5 - 5z^3 + 10z - \frac{10}{z} + \frac{5}{z^3} - \frac{1}{z^5}$$

   $$32i\sin^5\theta = \left(z^5 - \frac{1}{z^5}\right) - 5\left(z^3 - \frac{1}{z^3}\right) + 10\left(z - \frac{1}{z}\right)$$

   $$32i\sin^5\theta = 2i\sin 5\theta - 5(2i\sin 3\theta) + 10(2i\sin\theta)$$

   $$\sin^5\theta = \frac{1}{16}(\sin 5\theta - 5\sin 3\theta + 10\sin\theta)$$

   Thus $\displaystyle\int(10\sin\theta - 16\sin^5\theta)\,d\theta = \int(5\sin 3\theta - \sin 5\theta)\,d\theta$

   $$= -\frac{5}{3}\cos 3\theta + \frac{1}{5}\cos 5\theta + c$$

3. Prove $\cos^4 \theta + \sin^4 \theta = \dfrac{1}{4}(\cos 4\theta + 3)$

$z + \dfrac{1}{z} = 2\cos\theta, \quad z - \dfrac{1}{z} = 2i\sin\theta \text{ and } z^n + \dfrac{1}{z^n} = 2\cos n\theta$

$\left(z + \dfrac{1}{z}\right)^4 = 16\cos^4\theta = z^4 + 4z^3\dfrac{1}{z} + 6z^2\dfrac{1}{z^2} + 4z\dfrac{1}{z^3} + \dfrac{1}{z^4}$

$\qquad\qquad = \left(z^4 + \dfrac{1}{z^4}\right) + 4\left(z^2 + \dfrac{1}{z^2}\right) + 6$

Thus $16\cos^4\theta = 2\cos 4\theta + 8\cos 2\theta + 6$ $\qquad\qquad\qquad$ (1)

Similarly, $\left(z - \dfrac{1}{z}\right)^4 = 16i^4\sin^4\theta = \left(z^4 + \dfrac{1}{z^4}\right) - 4\left(z^2 + \dfrac{1}{z^2}\right) + 6$

$\qquad\qquad\qquad = 2\cos 4\theta - 8\cos 2\theta + 6$ $\qquad$ (2)

(1) + (2) gives $16(\cos^4\theta + \sin^4\theta) = 4\cos 4\theta + 12$

$\qquad$ and $\cos^4\theta + \sin^4\theta = \dfrac{1}{4}(\cos 4\theta + 3)$

4. Find, using De Moivre's theorem, $\displaystyle\int_0^{\pi/4} 8\cos^4\theta\, d\theta$

Using (1) above $\displaystyle\int_0^{\pi/4} 8\cos^4\theta\, d\theta = \int_0^{\pi/4} \cos 4\theta + 4\cos 2\theta + 3\, d\theta$

$\qquad\qquad\qquad = \left[\dfrac{1}{4}\sin 4\theta + 2\sin 2\theta + 3\theta\right]_0^{\pi/4}$

$\qquad\qquad\qquad = \left(\dfrac{1}{4}\sin\pi + 2\sin\dfrac{\pi}{2} + \dfrac{3\pi}{4}\right) - (0)$

$\qquad\qquad\qquad = 2 + \dfrac{3\pi}{4}$

13 The cube roots of unity

You have already met (Chapter 2) the roots of $z^3 = 1$ found by a basic algebraic method, and were invited to think about the nth roots of unity. Now we will explore further.

$z^3 = 1$ gives roots

$$z_1 = 1, \quad z_2 = -\frac{1}{2} + \frac{\sqrt{3}}{2}i, \quad z_3 = -\frac{1}{2} - \frac{\sqrt{3}}{2}i$$

or

$$1, \quad \cos\frac{2\pi}{3} + i\sin\frac{2\pi}{3}, \quad \cos\frac{4\pi}{3} + i\sin\frac{4\pi}{3}$$

Now, drawing an Argand diagram, it can be seen that these three roots lie on a unit circle, centre the origin, equally spaced around it, with an angle of $2\pi/3$ between each vector $\mathbf{z}$.

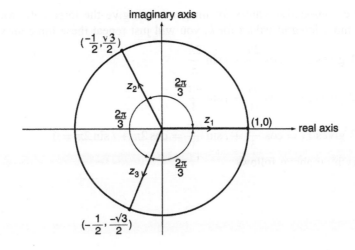

If these three roots are called 1, w, w^2, where w and w^2 are the two complex roots, clearly if $w = \cos(2\pi/3) + i\sin(2\pi/3)$, it is obvious by De Moivre's theorem that the other,

$$\cos\frac{4\pi}{3} + i\sin\frac{4\pi}{3} = \left(\cos\frac{2\pi}{3} + i\sin\frac{2\pi}{3}\right)^2 = w^2$$

It is not obvious (but equally true), that if the root

$$\left(\cos\frac{4\pi}{3} + i\sin\frac{4\pi}{3}\right) = w, \text{ then } \left(\cos\frac{4\pi}{3} + i\sin\frac{4\pi}{3}\right)^2$$

$$= \cos\frac{8\pi}{3} + i\sin\frac{8\pi}{3}$$

$$= \cos\frac{2\pi}{3} + i\sin\frac{2\pi}{3} = w^2$$

The three cube roots of unity can be written 1, w, w^2.
Therefore, the three cube roots of unity can be written 1, w, w^2, where $1 + w + w^2 = 0$, and $w^3 = 1$, and if one of the complex roots is w the other is w^2. These cube roots can also be found using De Moivre's theorem:

$z^3 = 1$, and 1 can be rewritten as a general complex number:

$$1 = \cos 2k\pi + i\sin 2k\pi \qquad\qquad k \text{ integer}$$

$$z^3 = \cos 2k\pi + i\sin 2k\pi$$

$$z = (\cos 2k\pi + i\sin 2k\pi)^{1/3}$$

$$= \cos\frac{2k\pi}{3} + i\sin\frac{2k\pi}{3} \qquad \text{by De Moivre's theorem.}$$

Any three consecutive values for integer k will give the three cube roots. If you use more integral values for k, you will just repeat these three answers:

$$k = 1 \text{ gives } z_1 = \cos\frac{2\pi}{3} + i\sin\frac{2\pi}{3}$$

$$k = 2 \text{ gives } z_2 = \cos\frac{4\pi}{3} + i\sin\frac{4\pi}{3}$$

$$k = 3 \text{ gives } z_3 = \cos\frac{6\pi}{3} + i\sin\frac{6\pi}{3} = \cos 2\pi + i\sin 2\pi = 1$$

and everything above follows.

Therefore, to summarize, and because these results are important and often required:

when $z^3 = 1$, the three roots are:

1. $z_1 = 1$

$$z_2 = \frac{1}{2} + \frac{\sqrt{3}}{2}i = \cos\frac{2\pi}{3} + i\sin\frac{2\pi}{3}$$

$$z_3 = -\frac{1}{2} - \frac{\sqrt{3}}{2}i = \cos\frac{4\pi}{3} + i\sin\frac{4\pi}{3} = \cos\left(-\frac{2\pi}{3}\right) + i\sin\left(-\frac{2\pi}{3}\right)$$

2. $(z_2)^2 = z_3$ and $(z_3)^2 = z_2$.
3. If one of the complex roots is denoted by w, the other is w^2.
4. The three cube roots of unity are often denoted by 1, w, w^2.
5. The vectors z_1, z_2, z_3 form an equilateral triangle and so

$$\mathbf{z}_1 + \mathbf{z}_2 + \mathbf{z}_3 = 0$$

thus $1 + w + w^2 = 0$

$$\left(N.B.\ 1 + \left(-\frac{1}{2} + \frac{\sqrt{3}}{2}i\right) + \left(-\frac{1}{2} - \frac{\sqrt{3}}{2}i\right) = 0 \right)$$

6. z_2 and z_3 are complex conjugates.
7. w and w^2 are complex conjugates.
8. $w^3 = 1$.

Simplify $(w^3 + w^4 + w^5)^2$ when w is a complex cube root of unity. Using $w^3 = 1$

$$(w^3 + w^4 + w^5)^2 = (1 + w + w^2)^2 = 0$$

Exercise 13.1

1. If w is a complex cube root of unity, simplify:

    (a) $(1 + 3w)(1 + 3w^2)$

    (b) $(1 + 3w + w^2)^2$

    (c) $(1 + w + 3w^2)^2$

    (d) Show that the product of (b) and (c) is equal to 16, and their sum is -4.

2. If $w = \cos(2\pi/3) + i\sin(2\pi/3)$, represent the numbers w, w^2 and w^3 in an Argand diagram, and find all possible values of $(w^n + w^{2n} + w^{3n})$, where n is an integer.

Exercise 13.1

Worked answers

1. (a)　$(1 + 3w)(1 + 3w^2) = 1 + 3w + 3w^2 + 9w^3$

$$= 1 + 3w + 3w^2 + 9 \qquad \text{(using } w^3 = 1)$$

$$= 3 + 3w + 3w^2 + 7$$

$$= 7 \qquad \text{(using } 1 + w + w^2 = 0)$$

(b)　$(1 + 3w + w^2)^2 = (1 + w + w^2 + 2w)^2$

$$= (2w)^2 \qquad \text{(using } 1 + w + w^2 = 0)$$

$$= 4w^2$$

(c)　$(1 + w + 3w^2)^2 = (1 + w + w^2 + 2w^2)^2$

$$= (2w^2)^2 \qquad \text{(using } 1 + w + w^2 = 0)$$

$$= 4w^4 = 4w \qquad \text{(using } w^3 = 1)$$

(d)　$4w^2 \times 4w = 16w^3 = 16 \qquad \text{(using } w^3 = 1)$

and $4w^2 + 4w = (4 + 4w + 4w^2 - 4) = -4$

$$\text{(using } 1 + w + w^2 = 0)$$

2. $w = \cos\dfrac{2\pi}{3} + i\sin\dfrac{2\pi}{3}$

$w^2 = \left(\cos\dfrac{2\pi}{3} + i\sin\dfrac{2\pi}{3}\right)^2$

$\qquad = \cos\dfrac{4\pi}{3} + i\sin\dfrac{4\pi}{3}$

$\qquad = \cos\left(-\dfrac{2\pi}{3}\right) + i\sin\left(-\dfrac{2\pi}{3}\right)$

$w^3 = \left(\cos\dfrac{2\pi}{3} + i\sin\dfrac{2\pi}{3}\right)^3$

$\qquad = \cos 2\pi + i\sin 2\pi = 1$

w, w^2 and w^3 are represented on an Argand diagram, by points on the unit circle as shown.

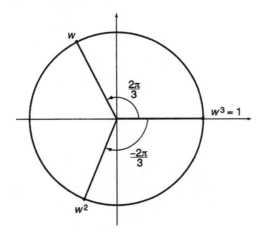

n integer: to find $w^n + w^{2n} + w^{3n}$
if n is multiple of 3, w^n, w^{2n} and w^{3n} will all equal 1 because $w^3 = 1$
$w^n + w^{2n} + w^{3n} = 1 + 1 + 1 = 3$
or if n is not a multiple of 3, $w^{3n} = 1$ and w^n and w^{2n} are equal to w
and w^2 or w^2 and w in every case
thus $w^n + w^{2n} + w^{3n} = w + w^2 + 1 = 0$

14 The *n*th roots of unity, where *n* is a positive integer

Clearly, the technique on page 72 is valid for **any** positive integer n, e.g. when $n = 8$, i.e. $z^8 = 1$, find the values of z.

$$z^8 = 1 = \cos 2k\pi + i \sin 2k\pi, k \text{ an integer}$$

$$z = (\cos 2k\pi + i \sin 2k\pi)^{1/8}$$

$$= \cos \frac{2k\pi}{8} + i \sin \frac{2k\pi}{8} \quad \text{by De Moivre's theorem}$$

$k = 1$ gives $z_1 = \cos \dfrac{2\pi}{8} + i \sin \dfrac{2\pi}{8} = \cos \dfrac{\pi}{4} + i \sin \dfrac{\pi}{4} = \dfrac{1}{\sqrt{2}} + \dfrac{1}{\sqrt{2}}$

$k = 2$ gives $z_2 = \cos \dfrac{\pi}{2} + i \sin \dfrac{\pi}{2} = i$

$k = 3$ gives $z_3 = \cos \dfrac{3\pi}{4} + i \sin \dfrac{3\pi}{4} = -\dfrac{1}{\sqrt{2}} + \dfrac{i}{\sqrt{2}}$

$k = 4$ gives $z_4 = \cos \pi + i \sin \pi = -1$

$k = 5$ gives $z_5 = \cos \dfrac{5\pi}{4} + i \sin \dfrac{5\pi}{4} = -\dfrac{1}{\sqrt{2}} - \dfrac{i}{\sqrt{2}}$

$k = 6$ gives $z_6 = \cos \dfrac{3\pi}{2} + i \sin \dfrac{3\pi}{2} = -i$

$k = 7$ gives $z_7 = \cos \dfrac{7\pi}{4} + i \sin \dfrac{7\pi}{4} = \dfrac{1}{\sqrt{2}} - \dfrac{i}{\sqrt{2}}$

$k = 8$ gives $z_8 = \cos 2\pi + i \sin 2\pi = 1$

and you can see that, as before, these roots are represented on the Argand diagram by points evenly spaced around the unit circle, this time dividing the central angle into segments of $\pi/4$.

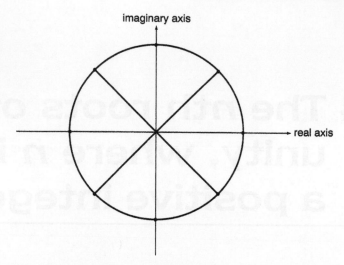

So the nth roots of unity are always around the unit circle, dividing it into n equal parts, where n is a positive integer.

By now you should be thinking, why does it have to be 'unity'?

Surely this should work for **any** number?

Yes, of course it does—see Chapter 15.

What about representing the roots on the Argand diagram? This will also be thought about in Chapter 15.

What about the modulus *of these numbers*? See Chapter 15.

By now you should also be thinking that there might be, **ought to be**, a better, briefer way of writing $\cos(5\pi/3) + i\sin(5\pi/3)$, for example!

Yes, there is – see Chapter 16.

Exercise 14.1

1. Find (a) the six roots of $z^6 = 1$, $\left.\right\}$ and show them all on
         (b) the six roots of $z^6 = 64$, $\left.\right\}$ an Argand diagram.

Exercise 14.1

Worked answers

1 (a) $z^6 = 1 = \cos 2\pi + i \sin 2\pi = \cos 2k\pi + i \sin 2k\pi$, k an integer

$$z = (\cos 2k\pi + i \sin 2k\pi)^{1/6} = \cos \frac{2k\pi}{6_3} + i \sin \frac{2k\pi}{6_3}$$

By De Moivre's theorem and six consecutive integral values for k will give the six roots. More integral values for k will just repeat these roots.

The roots are $z_1 = \cos \dfrac{\pi}{3} + i \sin \dfrac{\pi}{3}$ $(k = 1)$

$z_2 = \cos \dfrac{2\pi}{3} + i \sin \dfrac{2\pi}{3}$ $(k = 2)$

$z_3 = \cos \pi + i \sin \pi = -1$ $(k = 3)$

$z_4 = \cos \dfrac{4\pi}{3} + i \sin \dfrac{4\pi}{3} = \cos \left(-\dfrac{2\pi}{3} \right)$

$+ i \sin \left(-\dfrac{2\pi}{3} \right)$ $(k = 4 \ (or \ -2))$

$z_5 = \cos \dfrac{5\pi}{3} + i \sin \dfrac{5\pi}{3} = \cos \left(-\dfrac{\pi}{3} \right)$

$+ i \sin \left(-\dfrac{\pi}{3} \right)$ $(k = 5 \ (or \ -1))$

$z_6 = \cos 2\pi + i \sin 2\pi = \cos 0 + i \sin 0$

$= 1$ $(k = 6 \ (or \ 0))$

N.B. z_4 is $\overline{z_2}$ and $z_5 = \overline{z_1}$

(b) $z^6 = 64$ also has 6 roots, with the same arguments as above, using exactly the same theory, but each with modulus = 2. These roots are represented, on the diagram, by points z'

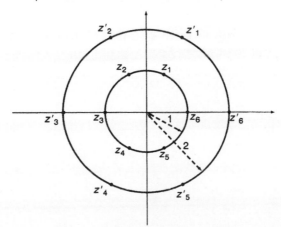

15 The *n*th roots of any complex number

This will work **exactly** as for Chapter 14, provided **you** can rewrite the number in a general r, θ form. If you cannot, refer back to Chapter 13 for more practice!

Now, for example, if $z^4 = -8 + 8\sqrt{3}i$, find the four values of z and illustrate them on an Argand diagram.

$$-8 + 8\sqrt{3}i = 16\left(-\frac{1}{2} + \frac{\sqrt{3}i}{2}\right) = 16\left(\cos\frac{2\pi}{3} + i\sin\frac{2\pi}{3}\right)$$

In general, $z^4 = 16\left(\cos\left(2n\pi + \frac{2\pi}{3}\right) + i\sin\left(2n\pi + \frac{2\pi}{3}\right)\right)$

thus $z = 2\left(\cos\left(2n\pi + \frac{2\pi}{3}\right) + i\sin\left(2n\pi + \frac{2\pi}{3}\right)\right)^{1/4}$

$$= 2\left(\cos 2\pi\frac{(3n+1)}{3} + i\sin 2\pi\frac{(3n+1)}{3}\right)^{1/4}$$

$$= 2\left(\cos\pi\frac{(3n+1)}{6} + i\sin\pi\frac{(3n+1)}{6}\right)$$

by De Moivre's theorem

$$z_1 = 2\left(\cos\frac{2\pi}{3} + i\sin\frac{2\pi}{3}\right):$$

$$z_2 = 2\left(\cos\frac{7\pi}{6} + i\sin\frac{7\pi}{6}\right)$$

$$z_3 = 2\left(\cos\frac{5\pi}{3} + i\sin\frac{5\pi}{3}\right):$$

$$z_4 = 2\left(\cos\frac{13\pi}{6} + i\sin\frac{13\pi}{6}\right) = 2\left(\cos\frac{\pi}{6} + i\sin\frac{\pi}{6}\right)$$

Clearly these are all to be found on a circle radius 2, centre the origin, positioned evenly around the circle: in this case, because there are four roots, they are positioned around the circle at angles $2\pi/4 = \pi/2$ apart — see diagram below.

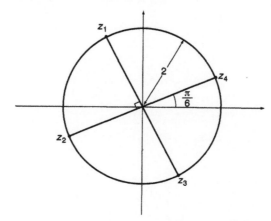

Exercise 15.1

1. Find the cube roots of -1 in the form $a + ib$, where a and b are real. Hence, solve the equation:

$$(z - 1)^3 + z^3 = 0$$

2. Given that $z^2 = \dfrac{20\sqrt{3} - 12i}{\sqrt{3} - 9i}$, find the two values of z, in the form

$$r(\cos\theta + i\sin\theta), \quad \text{where } r > 0, \text{ and } -\pi < \theta \le \pi$$

3. Find the four fourth roots of -4 in the form $a + ib$, where a and b are real. Plot the points corresponding to these roots on an Argand diagram.
   Hence, solve the equation:

$$(z + 1)^4 + 4(z - 1)^4 = 0$$

giving your answers in the form $c + id$, where c and d are real.

Exercise 15.1

Worked answers

1. $z^3 = -1 = \cos\pi + i\sin\pi = \cos(2k+1)\pi + i\sin(2k+1)\pi$ k integer

$$z = ((\cos(2k+1)\pi + i\sin(2k+1)\pi)^{1/3}$$

$$= \cos\left(\frac{2k+1}{3}\right)\pi + i\sin\left(\frac{2k+1}{3}\right)\pi \quad \text{by De Moivre's theorem}$$

$$z_1 = \cos\pi + i\sin\pi = -1 \quad (k=1)$$

$$z_2 = \cos\frac{\pi}{3} + i\sin\frac{\pi}{3} = \frac{1}{2} + i\frac{\sqrt{3}}{2} \quad (k=0)$$

$$z_3 = \cos\left(-\frac{\pi}{3}\right) + i\sin\left(-\frac{\pi}{3}\right) = \frac{1}{2} - i\frac{\sqrt{3}}{2} \quad (k=-1)$$

roots are $-1, \dfrac{1}{2} + i\dfrac{\sqrt{3}}{2}, \dfrac{1}{2} - i\dfrac{\sqrt{3}}{2}$ \hfill (1)

Hence to solve $(z-1)^3 + z^3 = 0$

i.e. $(z-1)^3 = -z^3 \Rightarrow \left(\dfrac{z-1}{z}\right)^3 = -1$

$$\frac{z-1}{z} = -1, \frac{1}{2} + i\frac{\sqrt{3}}{2}, \frac{1}{2} - i\frac{\sqrt{3}}{2} \quad \text{(using (1) above)}$$

$$1 - \frac{1}{z} = -1, \frac{1}{2} + i\frac{\sqrt{3}}{2}, \frac{1}{2} - i\frac{\sqrt{3}}{2}$$

$$-\frac{1}{z} = -2, -\frac{1}{2} + i\frac{\sqrt{3}}{2}, -\frac{1}{2} - i\frac{\sqrt{3}}{2}$$

$$\frac{1}{z} = 2, \frac{1}{2} - i\frac{\sqrt{3}}{2}, \frac{1}{2} + i\frac{\sqrt{3}}{2}$$

$$z = \frac{1}{2}, \frac{2}{(1 - i\sqrt{3})}, \frac{2}{(1 + i\sqrt{3})}$$

$$= \frac{1}{2}, \frac{2(1 + i\sqrt{3})}{(1 - i\sqrt{3})(1 + i\sqrt{3})}, \frac{2(1 - i\sqrt{3})}{(1 - i\sqrt{3})(1 + i\sqrt{3})}$$

$$= \frac{1}{2}, \frac{2(1 + i\sqrt{3})}{4}, \frac{2(1 - i\sqrt{3})}{4}$$

$$= \frac{1}{2}, \frac{1}{2}(1 + i\sqrt{3}), \frac{1}{2}(1 - i\sqrt{3})$$

2. $z^2 = \dfrac{20\sqrt{3} - 12i}{\sqrt{3} - 9i} = \dfrac{(20\sqrt{3} - 12i)(\sqrt{3} + 9i)}{(\sqrt{3} - 9i)(\sqrt{3} + 9i)}$

$$= \frac{60 - 12\sqrt{3}i + 180\sqrt{3}i + 108}{3 + 81}$$

$$= \frac{168^2 + 168^2\sqrt{3}i}{84_1}$$

$z^2 = 2 + 2\sqrt{3}i = (a + bi)^2$

$2 + 2\sqrt{3}i = a^2 - b^2 + 2abi$

$a^2 - b^2 = 2$ and $ab = \sqrt{3}$

$a^2 - \dfrac{3}{a^2} = 2 \Rightarrow a^4 - 2a^2 - 3 = 0 \Rightarrow (a^2 - 3)(a^2 + 1) = 0$

$a^2 = 3 \Rightarrow a = \pm\sqrt{3} \Rightarrow b = \pm 1$

$z = \sqrt{3} + i$ or $-\sqrt{3} - i$

$z = 2\left(\dfrac{\sqrt{3}}{2} + \dfrac{1}{2}i\right) = 2\left(\cos\dfrac{\pi}{6} + i\sin\dfrac{\pi}{6}\right)$

or $z = 2\left(-\dfrac{\sqrt{3}}{2} - \dfrac{1}{2}i\right) = 2\left(\cos\left(-\dfrac{5\pi}{6}\right) + i\sin\left(-\dfrac{5\pi}{6}\right)\right)$

3. $z^4 = -4 = 4(\cos(2k + 1)\pi + i\sin(2k + 1)\pi)$ (i.e. $4(-1)$)

$z = \sqrt{2}\left(\cos\left(\dfrac{2k+1}{4}\right)\pi + i\sin\left(\dfrac{2k+1}{4}\right)\pi\right)$

by De Moivre's theorem

when $k = 0$, $z = \sqrt{2}\left(\cos\dfrac{\pi}{4} + i\sin\dfrac{\pi}{4}\right) = \sqrt{2}\left(\dfrac{1}{\sqrt{2}} + \dfrac{i}{\sqrt{2}}\right) = 1 + i$

i.e. argument $= \dfrac{\pi}{4}$

Similarly, $k = 1$ gives $z = -1 + i$ \quad argument $= \dfrac{3\pi}{4}$

$k = 2$ gives $z = -1 - i$ \quad argument $= \dfrac{5\pi}{4}$

$k = 3$ gives $z = 1 - i$ \quad argument $= \dfrac{7\pi}{4}$

Thus the four fourth roots of -4 are $1 + i, 1 - i, -1 + i, -1 - i$.

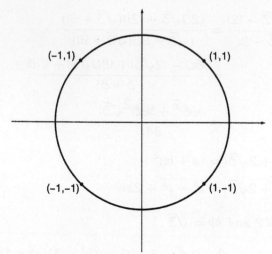

Solve $(z+1)^4 + 4(z-1)^4 = 0$

$$\left(\frac{z+1}{z-1}\right)^4 = -4$$

$$\frac{z+1}{z-1} = (1+i),\ (1-i),\ (-1+i),\ (-1-i)$$

Using the 4 fourth roots of -4 as already found

$$\frac{z+1}{z-1} = (1+i),\ (1-i),\ (-1+i),\ (-1-i)$$

Rewriting $\dfrac{z+1}{z-1}$

$$\frac{z+1}{z-1} = \frac{(z-1)+2}{z-1} = \frac{z-1}{z-1} + \frac{2}{z-1} = 1 + \frac{2}{z-1}$$

Therefore (i) $1 + \dfrac{2}{z-1} = (1+i) \Rightarrow \dfrac{2}{z-1} = i$

(ii) $1 + \dfrac{2}{z-1} = (1-i) \Rightarrow \dfrac{2}{z-1} = -i$

(iii) $1 + \dfrac{2}{z-1} = (-1+i) \Rightarrow \dfrac{2}{z-1} = -2+i$

(iv) $1 + \dfrac{2}{z-1} = (-1-i) \Rightarrow \dfrac{2}{z-1} = -2-i$

Therefore (i) $\left.\begin{array}{l} \dfrac{2}{i} = z-1 \\[2mm] \text{(ii)}\ \dfrac{2}{-i} = z-1 \end{array}\right\}$ i.e. $z = \dfrac{2i}{\pm 1}$ i.e. $z = 1 \pm 2i$

(iii) $\left.\dfrac{2}{-2+i} = z-1 \right\}$

(ii) $\left.\dfrac{2}{-2-i} = z-1 \right\}$

i.e. $z = \dfrac{2}{-2 \pm i} + 1$

$$z = 1 + \dfrac{2(-2 \mp i)}{(-2 \pm i)(-2 \mp i)}$$

$$= 1 + \dfrac{2}{5}(-2 \pm i)$$

$$\Rightarrow z = \dfrac{1}{5} \pm \dfrac{2}{5}i$$

Therefore the four roots are

$$z = 1 + 2i, \quad 1 - 2i, \quad \frac{1}{5}(1 + 2i), \quad \frac{1}{5}(1 - 2i)$$

16 The exponential form for a complex number

As a series, with θ measured in radians

$$\cos\theta = 1 - \frac{\theta^2}{2!} + \frac{\theta^4}{4!} - \frac{\theta^6}{6!} + \ldots \qquad (1)$$

$$\sin\theta = \theta - \frac{\theta^3}{3!} + \frac{\theta^5}{5!} - \ldots \qquad (2)$$

$$e^\theta = 1 + \theta + \frac{\theta^2}{2!} + \frac{\theta^3}{3!} + \frac{\theta^4}{4!} + \ldots \qquad (3)$$

We are interested in an alternative way of expressing $\cos\theta + i\sin\theta$ so $(1) + i(2)$ gives

$$\cos\theta + i\sin\theta = 1 + i\theta - \frac{\theta^2}{2!} - i\frac{\theta^3}{3!} + \frac{\theta^4}{4!} + i\frac{\theta^5}{5!} + \ldots$$

which can be written

$$= 1 + i\theta + \frac{(i\theta)^2}{2!} + \frac{(i\theta)^3}{3!} + \frac{(i\theta)^4}{4!} + \frac{(i\theta)^5}{5!} + \ldots$$

$$= e^{i\theta} \text{ by (3)}$$

Thus $\cos\theta + i\sin\theta = e^{i\theta}$, which allows Chapters 13, 14 and 15, for example, and any future work to be written more briefly. For example, page 83 $z = \cos(2k\pi/8) + i\sin(2k\pi/8)$ can become $e^{ik\pi/4}$ and z_1 to z_8 can be written $e^{i\pi/4}$, $e^{i\pi/2}$, $e^{3i\pi/4}$, etc.

So, any complex number z can now be written

$z = x + yi$

$\quad = r(\cos\theta + i\sin\theta) \qquad$ (θ in degrees or radians)

$\quad = re^{i\theta} \qquad$ (θ in radians)

and you use whichever of these forms you are asked for, or whichever seems the most appropriate for the particular question.

Example (a)

If $z_1 = 1 - i$ and $z_2 = \sqrt{3} + i$, convert z_1 and z_2 into the exponential form of a complex number and hence find $z_1 z_2$ and

$$\frac{z_1}{z_2} \text{ and } |z_1 z_2| \text{ and } \arg\left[\frac{z_1}{z_2}\right]$$

$z_1 = 1 - i = re^{i\theta}$ where $r = \sqrt{2}$ and $\theta = -(\pi/4)$. Plotting this point on the Argand diagram will, as ever, help you here.

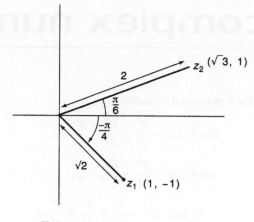

$$z_1 = \sqrt{2}e^{-i\pi/4}$$

$$z_2 = \sqrt{3} + i = re^{i\theta} \text{ where } r = 2 \text{ and } \theta = \frac{\pi}{6}$$

thus $z_2 = 2e^{i\pi/6}$

and $z_1 z_2 = 2\sqrt{2}e^{i(\pi/6 - \pi/4)} = 2\sqrt{2}e^{(-\pi/12)}$

thus $|z_1 z_2| = 2\sqrt{2}$

and $\dfrac{z_1}{z_2} = \dfrac{\sqrt{2}e^{-i\pi/4}}{2e^{i\pi/6}} = \dfrac{\sqrt{2}}{2}e^{i(-(\pi/4)-(\pi/6))} = \dfrac{\sqrt{2}}{2}e^{-5\pi i/12}$

thus $\arg\left[\dfrac{z_1}{z_2}\right] = -\dfrac{5\pi}{12}$

Example (b)

If $z = \sqrt{3} - i = re^{i\theta}$, find r and θ, where $r > 0$ and $-\pi < \theta \le \pi$, and hence find z^9 and $\bar{z}^9$ in the form $a + bi$, a and b real.

$$z = \sqrt{3} - i = 2\left(\frac{\sqrt{3}}{2} - \frac{i}{2}\right) = 2e^{-i\pi/6} \text{ thus } r = 2, \theta = -\frac{\pi}{6}$$

$$z^9 = 2^9 e^{-9\pi i/6} = 2^9 e^{-3\pi i/2} = 2^9 i$$

$$\bar{z} = \sqrt{3} + i = 2e^{i\pi/6}$$

$$\bar{z}^9 = 2^9 e^{9\pi i/6} = 2^9 e^{3\pi i/2} = -2^9 i$$

Now you try some:

Exercise 16.1

1. Find, in the form $re^{i\theta}$, where $r > 0$ and $-\pi < \theta \le \pi$, the four fourth roots of $8(-1 + i\sqrt{3})$.

2. Prove that $(z^n - e^{i\theta})(z^n - e^{-i\theta}) = z^{2n} - 2z^n \cos\theta + 1$.

   Hence find the roots of the equation $z^6 - z^3\sqrt{2} + 1 = 0$, in the form $\cos\alpha + i\sin\alpha$, where $-\pi < \alpha \le \pi$.

3. Find the roots z_1, z_2 and z_3 of the equation $z^3 = 1 - i$, giving your answers in the form $re^{i\theta}$, where $r > 0$ and $-\pi < \theta \le \pi$. Plot the points corresponding to these roots on an Argand diagram. Using the substitution $z = \dfrac{\omega - 1}{\omega + 1}$, find, in terms of z_1, z_2 and z_3, the roots of the equation:

$$(\omega - 1)^3 - (1 - i)(\omega + 1)^3 = 0$$

4. Express $\sqrt{3} - i$ in the form $re^{i\theta}$, where $r > 0$ and $-\pi < \theta \le \pi$ and hence show that:

$$(\sqrt{3} - i)^n + (\sqrt{3} + i)^n = 2^{n+1} \cos\left[\frac{n\pi}{6}\right]$$

Exercise 16.1

Worked answers

1. $z^4 = -8 + 8\sqrt{3}i = 16\left(-\frac{1}{2} + \frac{\sqrt{3}}{2}i\right) = 16\left(\cos\frac{2\pi}{3} + i\sin\frac{2\pi}{3}\right)$

i.e. $z^4 = 16e^{2\pi i/3}$

$\qquad = 16e^{(2\pi k + (2\pi/3))i}$ (in general) k integer

$\qquad = 16e^{2\pi(k + (1/3))i}$

$\qquad = 16e^{2\pi((3k+1)/3)i}$

$\therefore z = 2e^{2\pi((3k+1)/12)i} = 2e^{\pi((3k+1)/6)i}$

For $r > 0$ and $-\pi < \theta < \pi$, the four consecutive integral values of k, giving the four roots of the equation are $k = -2, -1, 0, 1$,

$z = 2e^{-(5\pi/6)i}$ $(k = -2)$

$z = 2e^{-(\pi/3)i}$ $(k = -1)$

$z = 2e^{(\pi/6)i}$ $(k = 0)$

$z = 2e^{(2\pi/3)i}$ $(k = 1)$

2. To prove that $(z^n - e^{i\theta})(z^n - e^{-i\theta}) = z^{2n} - 2z^n\cos\theta + 1$ (1)

LHS $= (z^n - e^{i\theta})(z^n - e^{-i\theta}) = z^{2n} - z^n(e^{i\theta} + e^{-i\theta}) + 1$

and $e^{i\theta} = \cos\theta + i\sin\theta$ and therefore $e^{i\theta} = \cos(-\theta) + i\sin(-\theta) = \cos\theta - i\sin\theta$

Therefore $e^{i\theta} + e^{-i\theta} = \cos\theta + i\sin\theta + \cos\theta - i\sin\theta = 2\cos\theta$

LHS $= z^{2n} - z^n(e^{i\theta} + e^{-i\theta}) + 1 = z^{2n} - 2z^n\cos\theta + 1 =$ RHS

Hence, to find the roots of $z^6 - z^3\sqrt{2} + 1 = 0$,

RHS of (1) with $n = 3$ is $z^6 - 2z^3\cos\theta + 1$,

and with $\theta = \dfrac{\pi}{4}$ is $z^6 - 2z^3\cos\dfrac{\pi}{4} + 1 = z^6 - 2z^3 \cdot \dfrac{1}{\sqrt{2}} + 1$

$$= z^6 - z^3\sqrt{2} + 1$$

Using (1), $z^6 - z^3\sqrt{2} + 1 = (z^3 - e^{i\pi/4})(z^3 - e^{-i\pi/4}) = 0$

thus $z^3 = e^{i\pi/4}$ or $e^{-i\pi/4}$ i.e. $z^3 = e^{(2n+(1/4))\pi i}$ or $e^{(2n-(1/4))\pi i}$

$\qquad\qquad\qquad = e^{((8n+1)/4)\pi i}$ or $e^{(8n-1)/4)\pi i}$

$\therefore z = e^{((8n+1)/12)\pi i}$ or $e^{((3n-1)/12)\pi i}$

There are six roots, and, in the form $\cos\alpha + i\sin\alpha$, $-\pi < \alpha < \pi$, they are,

$n = 0$, so that $\quad z = e^{\pi i/12} \quad = \cos\dfrac{\pi}{12} + i\sin\dfrac{\pi}{12};$

$$z = e^{-\pi i/12} = \cos\left(-\dfrac{\pi}{12}\right) + i\sin\left(-\dfrac{\pi}{12}\right)$$

$n = 1$, so that $\quad z = e^{3\pi i/4} \quad = \cos\dfrac{3\pi}{4} + i\sin\dfrac{3\pi}{4};$

$$z = e^{7\pi i/12} = \cos\dfrac{7\pi}{12} + i\sin\dfrac{7\pi}{12}$$

$n = -1$, so that $\quad z = e^{-7\pi i/12} = \cos\left(-\dfrac{7\pi}{12}\right) + i\sin\left(-\dfrac{7\pi}{12}\right);$

$$z = e^{-3\pi i/4} = \cos\left(-\dfrac{3\pi}{4}\right) + i\sin\left(-\dfrac{3\pi}{4}\right)$$

3. $z^3 = 1 - i = \sqrt{2}\left(\dfrac{1}{\sqrt{2}} - \dfrac{1}{\sqrt{2}}i\right) = \sqrt{2}\left(\cos\left(-\dfrac{\pi}{4}\right) + i\sin\left(-\dfrac{\pi}{4}\right)\right)$

$$= \sqrt{2}e^{(-\pi/4)i}$$

$z^3 = \sqrt{2}e^{(2\pi n - (\pi/4))i} = e^{i\pi(8n-1)/4}$ $\quad$ In general with n integer

$z = 2^{1/6}e^{\pi i(8n-1)/12}$

$r > 0$, $-\pi < \theta < \pi$, gives

$z_1 = 2^{1/6}e^{-\pi i/12};\ z_2 = 2^{1/6}e^{7\pi i/12};\ z_3 = 2^{1/6}e^{-3\pi i/4}$

$n = 0 \qquad\qquad n = 1 \qquad\qquad n = -1$

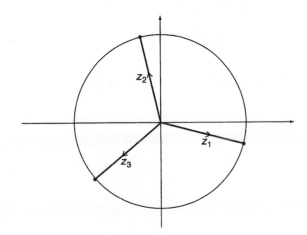

$$(w-1)^3 - (1-i)(w+1)^3 = 0 \text{ is } \left(\dfrac{w-1}{w+1}\right)^3 = 1 - i$$

and by putting $z = \dfrac{w-1}{w+1}$, the above results apply for $\dfrac{w-1}{w+1}$: i.e.

$$\frac{w-1}{w+1} = z_1, z_2 \text{ or } z_3$$

let $\dfrac{w-1}{w+1} = z_R$ where $R = 1, 2, 3$

then $w - 1 = z_R w + z_R$

and $w(1 - z_R) = 1 + z_R$

thus $w = \dfrac{1 + z_R}{1 - z_R}$ where $R = 1, 2, 3$ as in (1) on previous page.

4. $\sqrt{3} - i = 2\left(\dfrac{\sqrt{3}}{2} - \dfrac{1}{2}i\right) = 2\left(\cos\left(-\dfrac{\pi}{6}\right) + i\sin\left(-\dfrac{\pi}{6}\right)\right)$ and

$\cos\theta + i\sin\theta = e^{i\theta}$ thus $\sqrt{3} - i = 2e^{-\pi i/6}$

Hence to show that $(\sqrt{3} - i)^n + (\sqrt{3} + i)^n = 2^{n+1}\cos\dfrac{n\pi}{6}.$ (1)

$\sqrt{3} - i = 2e^{-\pi i/6}$ and similarly $\sqrt{3} + i = 2e^{\pi i/6}$

Since $e^{i\theta} = \cos\theta + i\sin\theta$, then $e^{-i\theta} = \cos(-\theta) + i\sin(-\theta) = \cos\theta - i\sin\theta$ and $e^{i\theta} + e^{-i\theta} = 2\cos\theta$ (2)

$\begin{aligned}
\text{LHS of (1)} &= (\sqrt{3} - i)^n + (\sqrt{3} + i)^n \\
&= (2e^{-\pi i/6})^n + (2e^{\pi i/6})^n \\
&= 2^n(e^{-\pi ni/6} + e^{\pi ni/6}) \\
\text{and by (2),} &= 2^n\left(2\cos\dfrac{n\pi}{6}\right) \\
&= 2^{n+1}\cos\dfrac{n\pi}{6} = \text{RHS}
\end{aligned}$

Thus $(\sqrt{3} - i)^n + (\sqrt{3} + i)^n = 2^{n+1}\cos\dfrac{n\pi}{6}$

17 Locus questions (plural is loci)

A point $P(x,y)$ represents the complex number $z = x + iy$ on the Argand diagram. By imposing special conditions on z, the positions of P will be controlled and will produce a locus for P.

(1) Loci involving $|z|$, the modulus of a complex number

First, really understand what the modulus of a complex number means, in visual terms, so that the locus of a point can be visualized on an Argand diagram.

For example $|x + iy|$ means, on an Argand diagram, the distance between the point (x,y) and the origin $(0,0)$.

If $z_1 = x_1 + iy_1$ and $z_2 = x_2 + iy_2$;

$$\text{then } |z_1 - z_2| = |(x_1 + iy_1) - (x_2 + iy_2)|$$

$$= |(x_1 - x_2) + i(y_1 - y_2)|$$

$$= \sqrt{(x_1 - x_2)^2 + (y_1 - y_2)^2} \text{ by definition}$$

which is the formula for the distance between the points (x_1,y_1) and (x_2,y_2) in 2-dimensional Cartesian coordinate geometry.

$$\text{Therefore } |z_1 - z_2| = \text{ distance between } (x_1,y_1) \text{ and } (x_2,y_2).$$

$$\text{and } |z - 3| = \text{ distance between } (x,y) \text{ and } (3,0).$$

$$\text{and } |z - 2i| = \text{ distance between } (x,y) \text{ and } (0,2).$$

Exercise 17.1

Try to translate the following moduli in exactly the same way.

1. $|z - 4|$ 2. $|z + 2|$ 3. $|z + i|$

4. $|z - (2 + i)|$ 5. $|z - 5 + 5i|$ 6. $|z + 2 - i|$

Now continue this translation to find, for example, the locus of the point z, which moves on the Argand diagram, so that $|z - 1| = 3$.

That is the distance between $P(x,y)$ and $(1,0)$ is 3. Think about what this means. It means that P is moving so that its distance from point $(1,0)$ is always 3 units, i.e.

$$P \text{ is moving on a circle centre } (1,0), \text{ radius } 3 \tag{1}$$

$$\text{i.e. circle } (x - 1)^2 + y^2 = 3^2 \tag{2}$$

Always read the question carefully, so that you answer precisely what is asked of you. Maybe what is required is a verbal description of the locus, i.e. (1) maybe a Cartesian equation, i.e. (2) maybe a sketch, maybe a shaded area, maybe a combination of these.

Try the following questions.

Exercise 17.2

Sketch the locus of the point $P(x,y)$ where P represents the complex number $z = x + iy$ on an Argand diagram, and write down the Cartesian equation of each locus, when:

1. $|z - 3| = 2$

2. $|z + i| = 1$

3. $|z - 1 + 2i| = 3$

Shade on an Argand diagram the areas represented by:

4. $|z - 2| < 3$

5. $|z + 1 - i| \leq 1$

6. $|z - i| > 2$

Sometimes there are other questions which arise involving locus and modulus problems. Common sense, thought and the use of what you have learned in this chapter so far will enable you to handle all these problems.

For example: Find the locus of the point z which moves so that $|z - 1| = |z + i|$.

(a) Translate this into words and you should find it quite simple.

$|z - 1| =$ distance between (x,y) and $(1,0)$.
$|z + i| =$ distance between (x,y) and $(0,-1)$.

(b) Put your translations on to a diagram.

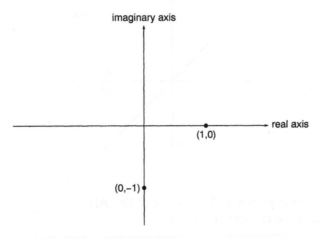

(c) Now $P(x,y)$ can move so that it is always equidistant from $(1,0)$ and $(0,-1)$. Here is where the common sense comes in. Think. Clearly P must lie on the line shown below.

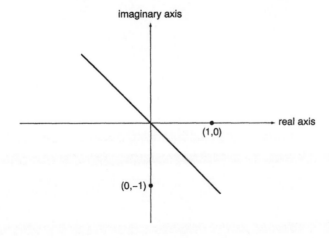

(d) Now, your problem is, how do they want this question answered? It is not very well worded. You must judge and decide. How many marks is it worth? How much time have you got? If you have the time, describe the locus in words, draw the diagram and give the Cartesian equation.

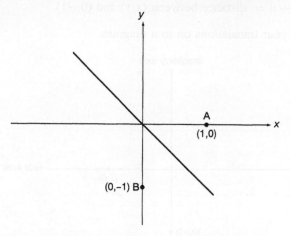

Locus is perpendicular bisector of line AB.

Equation of locus is $y = -x$.

Sometimes these techniques are not going to work for you, and the easiest way to find the required locus is to use the modulus definition (see next page).

For example: If P(x,y) is the point on the Argand diagram representing the complex number $z = x + iy$ and $|z - 2| = x + 2$, sketch the locus of P and express the equation of this locus in Cartesian form.

Translating this as before, the distance between (x,y) and $(2,0)$ is equal to $x + 2$, and this cannot be easily visualized.

Using the modulus definition:

$$|z - 2| = x + 2$$
$$\sqrt{(x - 2)^2 + y^2} = x + 2$$
$$(x - 2)^2 + y^2 = (x + 2)^2$$
$$x^2 - 4x + 4 + y^2 = x^2 + 4x + 4$$
$$y^2 = 8x$$

and this is a parabola (as shown).

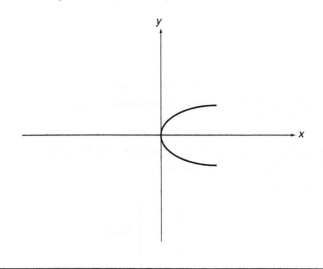

(2) Loci involving arg z, the argument of a complex number z

$\arg(z-1) = \theta$ translates to $\arg(x-1+iy) = \theta$, which means the angle between PA and the positive direction of the x-axis $= \theta$, where P is point (x,y) and A is $(1,0)$.

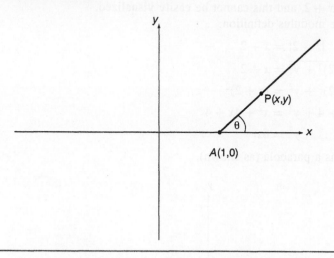

For example, sketch the locus represented by $\arg(z-a) = \pi/2$, where $a = 1+i$.

If $a = 1+i$, this is represented on the Argand diagram by point A(1,1) and P(x,y) moves so that the angle that PA makes with the positive direction of the x-axis is $\pi/2$.

Thus locus is

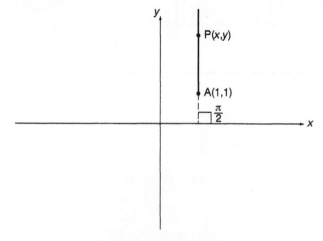

and the equation of the locus is the part-line $x = 1$, where $y \geq 1$.

Exercise 17.3

On the Argand diagram, sketch the locus represented by:

1. $\arg(z - i) = \pi/4$

2. $\arg \dfrac{z - 1}{z + 1} = \pi/2$

3. If $\arg \dfrac{z - 1}{z + 1} = \pi/4$, show that the locus of P(x,y), where $z = x + iy$ lies on the arc of a circle, and find the coordinates of the centre of the circle.

GENERAL LOCUS QUESTIONS

Exercise 17.4

(Advanced level questions)

1. Sketch on an Argand diagram the locus of a point P representing the complex number z, where

$$|z - 1| = |z - 3i|$$

and find z when $|z|$ has its least value on this locus.

2. On separate Argand diagrams represent, by shading, the regions where:

   (a) $|z| > 3$

   (b) $|z - 2| < |z - 4|$

   (c) $0 < \arg(z + 3) < \pi/6$ where $z = x + iy$.

   Hence (d) sketch the regions represented by (a), (b) and (c) simultaneously.

3. Find the ratio of the greatest value of $|z + 1|$ to its least value when $|z - i| = 1$.

Exercise 17.1

Worked answers

1. $|z - 4| = $ distance between (x,y) and $(4,0)$, on Argand diagram.

2. $|z + 2| = |z - (-2)| = $ distance between points (x,y) and $(-2,0)$.

3. $|z + i| = |z - (-i)| = $ distance between points (x,y) and $(0,-1)$.

4. $|z - (2 + i)| = $ distance between points (x,y) and $(2,1)$.

5. $|z - 5 + 5i| = |z - (5 - 5i)|$

   $= $ distance between points (x,y) and $(5,-5)$.

6. $|z + 2 - i| = |z - (-2 + i)|$

   $= $ distance between points (x,y) and $(-2,1)$.

Exercise 17.2

Worked answers

1. $|z - 3| = 2 \Rightarrow$ distance between (x,y) and $(3,0)$ is 2

   $\Rightarrow$ circle centre $(3,0)$, radius 2.

   $\Rightarrow (x - 3)^2 + y^2 = 2^2 \Rightarrow x^2 + y^2 - 6x + 5 = 0$

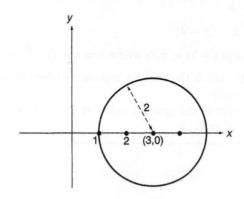

2. $|z + i| = 1 \Rightarrow$ distance between (x,y) and $(0,-1)$ is 1

        $\Rightarrow$ circle centre $(0,-1)$, radius 1.

        $\Rightarrow x^2 + (y+1)^2 = 1^2 \Rightarrow x^2 + y^2 + 2y = 0$

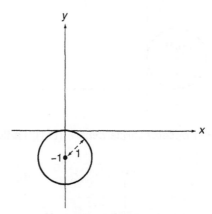

3. $|z - 1 + 2i| = 3 \Rightarrow$ distance between (x,y) and $(1,-2)$ is 3

        $\Rightarrow$ circle centre $(1,-2)$, radius 3.

        $\Rightarrow (x-1)^2 + (y+2)^2 = 3^2$

        $\Rightarrow x^2 + y^2 - 2x + 4y - 4 = 0$

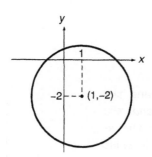

4. $|z - 2| < 3 \Rightarrow$ area inside circle centre $(2,0)$, radius 3.

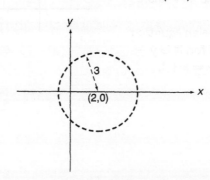

5. $|z + 1 - i| \leq 1 \Rightarrow |z - (-1 + i)| \leq 1 \Rightarrow$ area inside and including circle centre $(-1,1)$, radius 1.

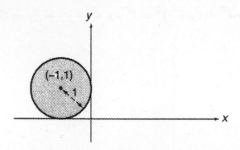

6. $|z - i| > 2 \Rightarrow$ area outside circle centre $(0,1)$, radius 2.

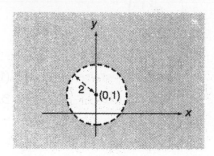

Exercise 17.3

Worked answers

1. $\arg(z - i) = \pi/4$ means the angle between PA and the positive direction of the x-axis is $\pi/4$, where P is point (x,y), and A is $(0,1)$.
Locus is part-line PA where $x \geq 0$ and $y \geq 1$.
Should the Cartesian equation of locus be asked for, it is $y - 1 = 1(x - 0) \Rightarrow y = x + 1$, where $x \geq 0$

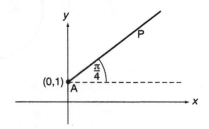

2. $\arg\left(\dfrac{z-1}{z+1}\right) = \dfrac{\pi}{2}$ where $z = x + yi$, represented by $P(x,y)$

$$\arg(z-1) - \arg(z+1) = \dfrac{\pi}{2}$$

$$\alpha \quad - \quad \beta \quad = \dfrac{\pi}{2}$$

where $\alpha = \arg(z-1) = $ angle between PA and positive direction of x-axis, where A is $(1,0)$

and $\beta = \arg(z+1) = $ angle between PB and positive direction of x-axis, where B is $(-1,0)$

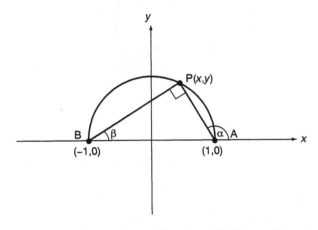

From the sketch, if $\alpha - \beta = \pi/2$, then $\angle APB = \pi/2$ and so P must lie on part of circle, diameter AB, centre $(0,0)$, radius 1, i.e. locus is the semicircle shown in the diagram.
Equation of locus is $x^2 + y^2 = 1$, $y \geq 0$.

3. $\arg\left(\dfrac{z-1}{z+1}\right) = \dfrac{\pi}{4}$.

As for question 2, the theory is similar, the final locus being an arc of a circle, because, this time, $\alpha - \beta = \pi/4$:

i.e. $\angle APB = \pi/4$ and then AB is subtending an angle $\pi/4$ at point P. See diagram.

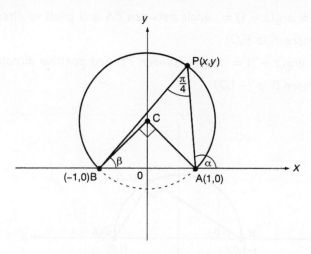

This circle is symmetrical about the y-axis, the centre C therefore being on the y-axis, and using the rule that ($\angle$ at centre = twice $\angle$ at edge subtended by same chord) $\angle ACB = 2\angle APB = \pi/2$, then using this symmetry, coordinates of C must be (0,1).

Exercise 17.4

Worked answers

1. $|z - 1| = |z - 3i|$

    distance between (x,y) and $A(1,0)$ = distance between (x,y) and $B(0,3)$

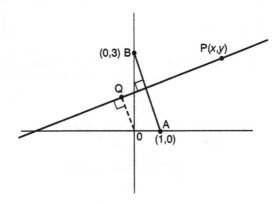

The required locus is a straight line, the perpendicular bisector of AB, and any point $P(x,y)$ lies on this line.

$|z|$ has its least values on this locus at point Q, where OQ is perpendicular to locus.

If Q has coordinates (a,b), to find $z = a + bi$.

To find where PQ and OQ meet.

OQ has gradient $-3/1$ and passes through O, thus OQ has equation $y = -3x$.

PQ has gradient $1/3$ and passes through $(1/2, 3/2)$ thus PQ has equation

$$y - \frac{3}{2} = \frac{1}{3}\left(x - \frac{1}{2}\right)$$

$$3y = x + 4$$

OQ, $y = -3x$ and PQ, $3y = x + 4$

meet when $3(-3x) = x + 4 \Rightarrow 10x = -4 \Rightarrow x = -\frac{2}{5}$ and $y = \frac{6}{5}$

thus Q is $\left(-\frac{2}{5}, \frac{6}{5}\right)$ and $z = -\frac{2}{5} + \frac{6}{5}i = \frac{1}{5}(-2 + 6i)$

2.

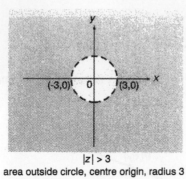

$|z| > 3$
area outside circle, centre origin, radius 3

(a)

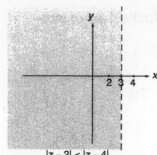

$|z - 2| < |z - 4|$
distance between (x,y) and $(2,0)$
< distance between (x,y) and $(4,0)$

(b)

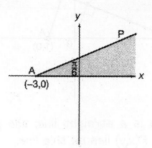

$0 < \arg(z+3) < \dfrac{\pi}{6}$
∠ between PA and Ax lies
between 0 and $\dfrac{\pi}{6}$

(c)

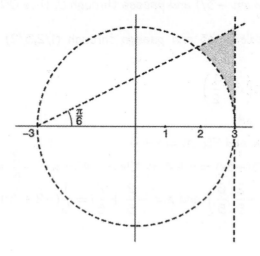

3. $|z - i| = 1$ means circle centre $(0,1)$, radius 1.

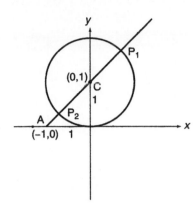

$|z + 1|$ means distance between $P(x,y)$ and $A(-1,0)$ where P is any point on the above circle.

Thus greatest value of $|z + 1| = AP_1 = AC + CP_1 = \sqrt{2} + 1$

least value of $|z + 1| = AP_2 = AC - CP_2 = \sqrt{2} - 1$

Then

$$\frac{\text{greatest}}{\text{least}} \text{ value of } |z + 1| = \frac{\sqrt{2} + 1}{\sqrt{2} - 1}$$

18 Transformations of the Argand diagram

Complex numbers can be used to describe transformations of a plane.

Example

If $z = x + iy$ is represented by the point $P(x,y)$ in the z-plane, and $\omega = u + iv$ is represented by the point $Q(u,v)$ in the ω-plane, then a relationship between z and ω will define a mapping of P to Q.

Examples

(i) If P describes the circle $x^2 + y^2 = 36$, and $\omega = z^2$ is the relationship between z and ω, then the locus of Q is found as follows:

   (a) $x^2 + y^2 = 36$ can be expressed in the form $|z| = 6$;
   (b) $\omega = z^2 \Rightarrow |\omega| = |z^2| = |z||z|$ and so $|z| = 6$ maps to $|\omega| = 36$;
   (c) the locus of Q is a circle centre $(0,0)$, radius 36 in the ω-plane.

(ii) If P describes the y-axis ($x = 0$), and the relationship between z and ω is $\omega = 1/(z + 1)$, then the locus of Q is found as follows: This time the link between z and ω is not so straightforward as in example (i), and so we must find u in terms of x and y, and v in terms of x and y, and hence $u + iv$ in terms of x and y: OR, as we need to find $z = x + iy$ in terms of u and v, and as $x = 0$ is the locus of P, then, instead of working out $\omega = 1/(z + 1)$, rearrange this relationship, to find z in terms of ω, i.e.

   (a) $\omega = \dfrac{1}{z+1} \Rightarrow z + 1 = \dfrac{1}{\omega} \Rightarrow z = \dfrac{1}{\omega} - 1$

   (b) $z = x + iy = \dfrac{1}{u+iv} - 1 = \dfrac{u - iv}{u^2 + v^2} - 1$

   (c) Since $x = 0$ is the locus of P, then $0 = \dfrac{u}{u^2 + v^2} - 1$

   (d) The locus of Q is $\dfrac{u}{u^2 + v^2} = 1$ or $u^2 + v^2 - u = 0$

Exercise 18.1

If $z = x + iy$ is represented by the point $P(x,y)$ on the z-plane, and $w = u + iv$ is represented by the point $Q(u,v)$ on the w-plane:

1. If $w = z^2$, find the image on the w-plane of the hyperbola $xy = 6$ on the z-plane, and illustrate these loci with sketches on the z-plane and the w-plane.

2. If $w = \dfrac{1}{z}$, find the image on the w-plane of $|z| = 8$.

3. If $w = \dfrac{1}{z}$, find the image on the w-plane of $\arg z = \dfrac{\pi}{4}$.

4. If $w = \dfrac{1}{z}$, find the image on the w-plane of $2y = 3x - 4$.

5. If $w = \dfrac{1 + zi}{z + i}$, (a) express u and v in terns of x and y.
   (b) Prove that when Q describes the portion of the imaginary axis between the points representing $-i$ and i, P describes the whole of the positive half of the imaginary axis.

Exercise 18.1

Worked answers

$z = x + iy$ represented on the Argand diagram by point $P(x,y)$ on the z-plane.

$w = u + iv$ represented on the Argand diagram by point $Q(u,v)$ on the w-plane.

1. If $w = z^2$, to find the transformation on the w-plane of the hyperbola $xy = 6$ on the z-plane.

$$u + iv = (x + iy)^2 = x^2 - y^2 + 2xyi$$

$v = 2xy$ and on the z-plane $xy = 6$

$v = 12$

on the w-plane, this transformation becomes the straight line $v = 12$.

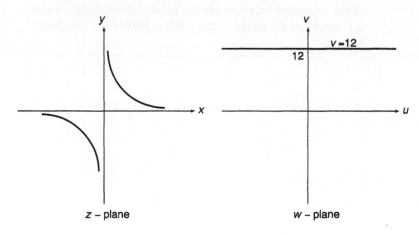

z – plane   w – plane

2. If $w = \dfrac{1}{z}$, to find the image on the w-plane of $|z| = 8$ on the z-plane.

This could be worked in a similar manner to Q1, but is more easily done by considering that, if $w = \dfrac{1}{z}$, then $|w| = \left|\dfrac{1}{z}\right| = \dfrac{1}{|z|}$ and when $|z| = 8$ on the z-plane, $|w| = \dfrac{1}{8}$.

$|z| = 8$ is a circle, centre origin, radius 8 on the z-plane,

thus $|w| = \dfrac{1}{8}$ is also a circle, centre origin, radius $\dfrac{1}{8}$ on the w-plane.

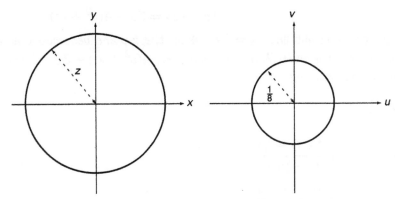

3. If $w = \dfrac{1}{z}$, to find the image on the w-plane of $\arg z = \dfrac{\pi}{4}$ on the z-plane.

If $w = \dfrac{1}{z}$, then $z = \dfrac{1}{w}$ and $\arg z = \arg \dfrac{1}{w}$

$$= \arg 1 - \arg w$$

$$\text{thus } \dfrac{\pi}{4} = 0 - \arg w$$

$\arg z = \dfrac{\pi}{4}$ maps onto $\arg w = -\dfrac{\pi}{4}$

i.e. when $P(x,y)$ lies on the part-line $y = x$, $x > 0$, then $Q(u,v)$ lies on the part-line $v = -u$, $u > 0$.

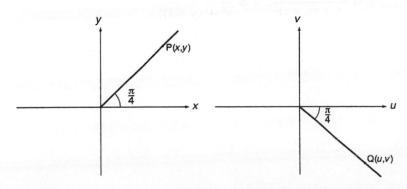

4. If $w = \dfrac{1}{z}$, to find the image on the w-plane of $2y = 3x - 4$ on the z-plane.

If $w = \dfrac{1}{z}$, then $z = \dfrac{1}{w}$ and $x + iy = \dfrac{1}{u + iv} = \dfrac{u - iv}{u^2 + v^2}$

$x = \dfrac{u}{u^2 + v^2}$ and $y = \dfrac{-v}{u^2 + v^2}$

$2y = 3x - 4$ is transformed to $\dfrac{-2v}{u^2 + v^2} = \dfrac{3u}{u^2 + v^2} - 4$

i.e. $-2v = 3u - 4(u^2 + v^2)$

Thus the straight line $2y = 3x - 4$ on the z-plane becomes transformed into the much more elaborate curve $4(u^2 + v^2) - 3u - 2v = 0$ under this mapping.

5. $w = \dfrac{1 + zi}{z + i}$

(a) $u + iv = \dfrac{1 + (x + iy)i}{(x + iy) + i}$

$= \dfrac{(1 - y) + xi}{x + i(y + 1)}$

$= \dfrac{((1 - y) + xi)(x - i(y + 1))}{(x + i(y + 1))(x - i(y + 1))}$

$= \dfrac{x(1 - y) + x(y + 1) + i(x^2 - (1 + y)(1 - y))}{x^2 + (y + 1)^2}$

$= \dfrac{2x + (x^2 + y^2 - 1)i}{x^2 + (y + 1)^2}$

Thus

$$u = \dfrac{2x}{x^2 + (y + 1)^2}, \quad v = \dfrac{x^2 + y^2 - 1}{x^2 + (y + 1)^2} \tag{1}$$

(b) To prove when Q describes the portion of the imaginary axis between $-i$ and i, i.e. when $u = 0$ and $-1 < v < 1$
then P describes the whole positive half of the imaginary axis, i.e. $x = 0$ and $y > 0$, i.e.
to prove that when $u = 0$ and $-1 < v < 1$, then $x = 0$ and $y > 0$

Using (1), when $u = 0$, $0 = \dfrac{2x}{x^2 + (y+1)^2}$, $\Rightarrow x = 0$

and thus $v = \dfrac{y^2 - 1}{(y+1)^2}$

$= \dfrac{(y-1)(y+1)}{(y+1)^2}$

and when $-1 < v < 1$, $-1 < \dfrac{y-1}{y+1} < 1$

LHS $-1 < \dfrac{y-1}{y+1} \Rightarrow 0 < \dfrac{y-1}{y+1} + 1 \Rightarrow 0 < \dfrac{y-1+y+1}{y+1}$

$$\Rightarrow 0 < \dfrac{2y}{y+1}$$

and this is valid provided $y > 0$ or $y < -1$.

RHS $\dfrac{y-1}{y+1} < 1 \Rightarrow \dfrac{y-1}{y+1} - 1 < 0 \Rightarrow \dfrac{y-1-(y+1)}{y+1} < 0$

$$\Rightarrow \dfrac{-2}{y+1} < 0,$$

and this is valid provided $y > -1$.
Thus LHS and RHS are both valid provided $y > 0$.
Therefore P describes the whole positive half of the imaginary axis.

19 A sample question

Pure Maths. A Level. London.

Q1. (i) The transformation $\omega = (z+2)/(z+i)$ where $z \neq -i$, $w \neq 1$, maps the complex number $z = x + iy$ onto the complex number

$$\omega = u + iv$$

(a) Show that, if the point representing ω lies on the real axis, the point representing z lies on a straight line.

(b) Show further that, if the point representing ω lies on the imaginary axis, then the point representing z lies on the circle:

$$\left| z + 1 + \frac{1}{2}i \right| = \frac{\sqrt{5}}{2}$$

(ii) Given that $2 + i$ is a root of the equation

$$x^4 + ax^3 + bx^2 + 25 = 0$$

where a and b are real, find the other three roots.

Worked answer

(i) (a) To show that, if ω lies on the real axis, i.e. if $\omega = u$ (i.e. $v = 0$) then z lies on a straight line.

$$\omega = \frac{z+2}{z+i}, \ z \neq -i, \ \omega \neq 1,$$

$$u + iv = \frac{x + iy + 2}{x + i(y+1)} = \frac{((x+2) + iy)(x - i(y+1))}{(x + i(y+1))(x - i(y+1))} \qquad (1)$$

Since $v = 0$, the imaginary part of $\omega = u + iv$, (1) equals zero.

i.e. $xy - (y+1)(x+2) = 0$

i.e. $xy - xy - x - 2 - 2y = 0$

$x + 2y + 2 = 0$, which is a straight line in the z-plane.

(b) If ω lies on the imaginary axis, i.e. if $u = 0$, then Real part of (1) equals zero.

i.e. $x(x+2) + y(y+1) = 0$

$x^2 + y^2 + 2x + y = 0$

$$(x+1)^2 + \left(y + \frac{1}{2}\right)^2 = 1 + \frac{1}{4}$$

i.e. circle centre $\left(-1, -\frac{1}{2}\right)$, radius $\frac{\sqrt{5}}{2}$

which is circle $\left|x + iy + 1 + \frac{1}{2}i\right| = \frac{\sqrt{5}}{2}$,

i.e. $\left|z + 1 + \frac{1}{2}i\right| = \frac{\sqrt{5}}{2}$

(ii) $(2 + i)$ is a root of $x^4 + ax^3 + bx^2 + 25 = 0$, a and b real   (2)

i.e. $x_1 = (2 + i)$ is a root $\Rightarrow x_2 = (2 - i)$ is another, and so the quadratic LHS contains

$(x^2 - (x_1 + x_2)x + (x_1 x_2))$ as a factor.

i.e. $(x^2 - 4x + 5)$

$x^4 + ax^3 + bx^2 + 25 = (x^2 - 4x + 5)(x^2 + cx + d)$, where $d = 5$.

Equating terms in x, $0 = -20 + 5c \Rightarrow c = 4$

then the remaining factor of LHS of (2) is $(x^2 + 4x + 5)$

so that (2) becomes $(x^2 - 4x + 5)(x^2 + 4x + 5) = 0$

and roots are $x = 2 + i, \ 2 - i \ \dfrac{-4 \pm \sqrt{(16 - 20)}}{2}$

$\Rightarrow x = 2 + i, \ 2 - i, \ -2 + i, \ -2 - i.$

The other three roots of (2) are $x = 2 - i, \ -2 + i, \ -2 - i.$

20 Further questions

Q1. Pure Maths. A Level. London.
(i) Given that $1 + 2i$ is a root of the equation

$$x^4 - 4x^3 - 6x^2 + 20x - 75 = 0$$

find the other three roots.

Plot the points representing all four of these roots on an Argand diagram. Hence, or otherwise, show that these points are the vertices of a rhombus with sides of length $2\sqrt{5}$.

(ii) Express $\sin^5 \theta$ in terms of sines of multiples of θ and hence evaluate:

$$\int_{\pi/6}^{\pi/2} \sin^5 \theta \, d\theta$$

Q2. Pure Maths. A Level. London.
(i) Points P and Q represent the complex numbers $z(= x + iy)$ and $\omega(= u + iv)$ in the z-plane and ω-plane, respectively. Given that z and ω are connected by the relation $\omega = \dfrac{z - i}{z + i}$ and that the locus of P is the x-axis, find the Cartesian equation of the locus of Q and sketch the locus of Q on an Argand diagram.

(ii) Find the cube roots of -1 in the form $a + ib$, where a and b are real. Hence, or otherwise, solve the equation:

$$(z - 1)^3 + z^3 = 0$$

Q3. Pure Maths. A Level. London.
(i) Given that $z^2 = \dfrac{20\sqrt{3} - 12i}{\sqrt{3} - 9i}$, find the two values of z, in the form:

$$r(\cos \theta + i \sin \theta), \quad \text{where } r > 0 \text{ and } -\pi < \theta \leq \pi$$

(ii) The transformation $T: z \longrightarrow \omega$ in the complex plane is defined by:
$$\omega = \frac{az + b}{z + c}, \quad \text{where } a, b, c \in R.$$
Given that $\omega = 3$, when $z = 0$ and that $\omega = 2 - i$ when $z = -1 + i$, find the values of the constants a, b and c.

Q4. Pure Maths. A Level. AEB.

Express $4(\sqrt{3} - i)$ in the form $re^{i\theta}$, where $r > 0$ and $-\pi < \theta \leq \pi$. (3 marks)

Find the cube roots z_1, z_2, z_3 of $4(\sqrt{3} - i)$ giving your answers in the form $re^{i\theta}$, where $r > 0$ and $-\pi < \theta \leq \pi$.   (5 marks)

Show on an Argand diagram the points 2 and $-2i$.   (1 mark)

Hence, or otherwise, show that if z satisfies the equation:

$$\arg\left(\frac{z-2}{z+2i}\right) = \frac{\pi}{4}$$

then the locus of z in the complex plane is an arc of a circle. State the radius and the centre of the circle. Show that just two of the points represented by z_1, z_2, z_3 on the Argand diagram lie on this arc of the circle.   (6 marks)

Q5. Pure Maths. A Level. AEB.

Giving your answers in the form $a + ib$, where a, $b \in R$, find the complex numbers u and v which satisfy the simultaneous equations:

$$3u + v = 4i$$

$$u - iv = 8 + 2i$$ (4 marks)

In the Argand diagram the point P represents the complex number

$$z = x + iy$$

Display the numbers u and v in an Argand diagram and show that the locus of all points P such that:

$$|z - u| = |z - v| \text{ is given by } 2y - x = 7$$ (5 marks)

Hence find the two complex numbers, $z = z_1$ and $z = z_2$, of this locus for which

$$|z - 4 - 3i| = 5$$ (4 marks)

Calculate, in radians, $\arg z_1$ and $\arg z_2$.   (2 marks)

Worked answers

Q1. London.

(i) $x^4 - 4x^3 - 6x^2 + 20x - 75 = 0$ has root $x_1 = 1 + 2i$.   (1)

To find the other three roots.

A second root is $x_2 = 1 - 2i$ (the complex conjugate of x_1)

$$(x - x_1)(x - x_2) = x^2 - (\text{sum of roots})x + (\text{product of roots})$$
$$= x^2 - 2x + 5 \text{ is a factor of LHS of (1)}$$

$x^4 - 4x^3 - 6x^2 + 20x - 75 = 0$ and factorizing LHS

$(x^2 - 2x + 5)(x^2 - 2x - 15) = 0$

$(x - x_1)(x - x_2)(x - 5)(x + 3) = 0$

Thus the other three roots of (1) are $x = 1 - 2i, 5, -3$.

To plot these points on an Argand diagram.

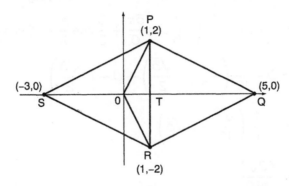

To show that these points are the vertices of a rhombus with sides $2\sqrt{5}$.

Clearly PQRS is a rhombus because its diagonals PR and QS bisect each other at right angles.

Thus four sides are equal, i.e. $PQ = QR = RS = SP$.

Considering $\triangle SPT$, $PT^2 + TS^2 = PS^2 \Rightarrow 2^2 + 4^2 = PS^2$
$$\Rightarrow PS^2 = 20 \Rightarrow PS = 2\sqrt{5}.$$

Thus PQRS is a rhombus with sides $2\sqrt{5}$.

(ii) $2i\sin\theta = z - \dfrac{1}{z}$ where $z = \cos\theta + i\sin\theta$ and $\dfrac{1}{z} = \cos\theta - i\sin\theta$

Similarly, $z^n - \dfrac{1}{z^n} = 2i\sin n\theta$ (using De Moivre's theorem)

$$(2i\sin\theta)^5 = \left(z - \frac{1}{z}\right)^5$$

and using Pascal's triangle and rearranging

$$= \left(z^5 - \frac{1}{z^5}\right) - 5\left(z^3 - \frac{1}{z^3}\right) + 10\left(z - \frac{1}{z}\right)$$

$$2^5 i^5 \sin^5\theta = 2i\sin 5\theta - 5(2i\sin 3\theta) + 10(2i\sin\theta)$$

and $i^5 = i^4 \cdot i = i$

$$\sin^5\theta = \frac{1}{16}(\sin 5\theta - 5\sin 3\theta + 10\sin\theta)$$

Hence $\displaystyle\int_{\pi/6}^{\pi/2} \sin^5\theta\, d\theta$

$$= \int_{\pi/6}^{\pi/2} \frac{1}{16}(\sin 5\theta - 5\sin 3\theta + 10\sin\theta)\, d\theta$$

$$= \frac{1}{16}\left[-\frac{1}{5}\cos 5\theta + \frac{5}{3}\cos 3\theta - 10\cos\theta\right]_{\pi/6}^{\pi/2}$$

$$= \frac{1}{16}(0) - \frac{1}{16}\left(-\frac{1}{5}\cos\frac{5\pi}{6} + \frac{5}{3}\cos\frac{\pi}{2} - 10\cos\frac{\pi}{6}\right)$$

$$= 0 - \frac{1}{16}\left(-\frac{1}{5}\left(-\frac{\sqrt{3}}{2}\right) + 0 - 10\frac{\sqrt{3}}{2}\right)$$

$$= -\frac{1}{16}\left(\frac{\sqrt{3}}{10} - 5\sqrt{3}\right)$$

$$= \frac{49\sqrt{3}}{160}$$

Q2. London.

Given $P(x,y)$ represents points on the z plane where $z = x + iy$
and $Q(u,v)$ represents points on the w plane where $w = u + iv$
and z and w are connected by the relation $w = \dfrac{z - i}{z + i}$

(i) $w = \dfrac{z - i}{z + i}$ thus $zw + iw = z - i$ and $z(1 - w) = i(1 + w)$

so that $z = i\dfrac{(1 + w)}{(1 - w)}$

Locus of P is x-axis — i.e. $y = 0$ — to find locus of Q.

Using $z = \dfrac{i(1 + w)}{(1 - w)}$ where $z = x + iy$ and $w = u + iv$

$x + iy = \dfrac{i(1 + u + iv)(1 - u + iv)}{(1 - u - iv)(1 - u + iv)}$ and since $y = 0$,

equating imaginary parts,

$0 = (1 + u)(1 - u) - v^2$

i.e. $u^2 + v^2 = 1$ is the locus of Q.

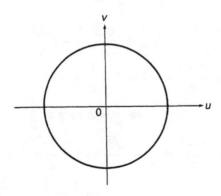

(ii) Let $z^3 = -1 = \cos(2n+1)\pi + i\sin(2n+1)\pi$ where n is integer

$\therefore z = (\cos(2n+1)\pi + i\sin(2n+1)\pi)^{1/3}$

$$= \cos\left(\frac{2n+1}{3}\right)\pi + i\sin\left(\frac{2n+1}{3}\right)\pi$$

using De Moivre's theorem,

and by choosing three consecutive integral values for n, the three roots can be found.

$n = 1$ gives $z = \cos\pi + i\sin\pi = -1$

$n = 0$ gives $z = \cos\dfrac{\pi}{3} + i\sin\dfrac{\pi}{3} = \dfrac{1}{2} + \dfrac{i\sqrt{3}}{2}$

$n = -1$ gives $z = \cos\left(-\dfrac{\pi}{3}\right) + i\sin\left(-\dfrac{\pi}{3}\right) = \dfrac{1}{2} - i\dfrac{\sqrt{3}}{2}$

and the cube roots of -1 are $-1, \dfrac{1}{2} + i\dfrac{\sqrt{3}}{2}, \dfrac{1}{2} - i\dfrac{\sqrt{3}}{2}.$ (1)

Hence, to solve $(z-1)^3 + z^3 = 0$

i.e. $\left(\dfrac{z-1}{z}\right)^3 = -1$

Using (1), $\dfrac{z-1}{z} = -1, \dfrac{1}{2} + i\dfrac{\sqrt{3}}{2}, \dfrac{1}{2} - i\dfrac{\sqrt{3}}{2}$

Thus either $\dfrac{z-1}{z} = -1 \Rightarrow 1 - \dfrac{1}{z} = -1$

$$\Rightarrow \frac{1}{z} = 2 \Rightarrow z = \frac{1}{2}$$

or $\dfrac{z-1}{z} = \dfrac{1}{2} + i\dfrac{\sqrt{3}}{2} \Rightarrow 1 - \dfrac{1}{z} = \dfrac{1}{2} + i\dfrac{\sqrt{3}}{2}$

$$\Rightarrow \frac{1}{z} = \frac{1}{2} - i\frac{\sqrt{3}}{2}$$

and $z = \dfrac{2}{1 - i\sqrt{3}} = \dfrac{2(1 + i\sqrt{3})}{(1 - i\sqrt{3})(1 + i\sqrt{3})}$

$$= \frac{2(1 + i\sqrt{3})}{1+3} = \frac{1}{2} + i\frac{\sqrt{3}}{2}$$

or similarly, $z = \dfrac{1}{2} - i\dfrac{\sqrt{3}}{2}.$

Thus the required solutions of $(z-1)^3 + z^3 = 0$ are

$$z = \frac{1}{2}, \frac{1}{2} + i\frac{\sqrt{3}}{2}, \frac{1}{2} - i\frac{\sqrt{3}}{2}.$$

Q3. London.

(i) $z^2 = \dfrac{20\sqrt{3} - 12i}{\sqrt{3} - 9i} = 4\dfrac{(5\sqrt{3} - 3i)(\sqrt{3} + 9i)}{(\sqrt{3} - 9i)(\sqrt{3} + 9i)}$

$\qquad = 4\dfrac{(15 + 27 - 3\sqrt{3}i + 45\sqrt{3}i)}{3 + 81}$

$\qquad = \dfrac{4}{84}(42 + 42\sqrt{3}i)$

$\qquad = 4\left(\dfrac{1}{2} + i\dfrac{\sqrt{3}}{2}\right)$

$\qquad = 4\left(\cos\dfrac{\pi}{3} + i\sin\dfrac{\pi}{3}\right)$

$\qquad z = \pm 2\left(\cos\dfrac{\pi}{6} + i\sin\dfrac{\pi}{6}\right)$

$\therefore z = 2\left(\cos\dfrac{\pi}{6} + i\sin\dfrac{\pi}{6}\right)$ and $2\left(-\cos\dfrac{\pi}{6} - i\sin\dfrac{\pi}{6}\right)$

$\qquad = 2\left(\cos\dfrac{7\pi}{6} + i\sin\dfrac{7\pi}{6}\right)$

$\qquad = 2\left(\cos\left(-\dfrac{5\pi}{6}\right) + i\sin\left(-\dfrac{5\pi}{6}\right)\right)$

$\therefore z_1 = 2\left(\cos\dfrac{\pi}{6} + i\sin\dfrac{\pi}{6}\right),$

$z_2 = 2\left(\cos\left(-\dfrac{5\pi}{6}\right) + i\sin\left(-\dfrac{5\pi}{6}\right)\right)$
where $r > 0$ and $-\pi < \theta \le \pi$

(ii) T: $z \to w$ where $w = \dfrac{az + b}{z + c}$, $a, b, c \in R$

when $w = 3$, $z = 0$ and therefore $3 = \dfrac{b}{c} \Rightarrow b = 3c$ \qquad (1)

Rearranging $w = \dfrac{az + b}{z + c}$ to make z the subject of the formula,

$wz + cw = az + b$ and so $z = \dfrac{b - cw}{w - a}$

when $z = -1 + i$, $w = 2 - i$ and therefore $-1 + i = \dfrac{b - c(2 - i)}{2 - i - a}$

Substituting (1), $-1 + i = \dfrac{c + ci}{2 - a - i} = \dfrac{c(1 + i)((2 - a) + i)}{((2 - a) - i)((2 - a) + i)}$

$-1 + i = \dfrac{c(2 - a - 1) + ci(2 - a + 1)}{(2 - a)^2 + 1} = \dfrac{c(1 - a) + ci(3 - a)}{(2 - a)^2 + 1}$

$-1 = \dfrac{c(1 - a)}{(2 - a)^2 + 1}$ (2)   and $1 = \dfrac{c(3 - a)}{(2 - a)^2 + 1}$ (3)

$\dfrac{(2)}{(3)}$ gives $\;-\dfrac{1}{1} = \dfrac{\cancel{c}(1 - a)}{\cancel{c}(3 - a)} \Rightarrow -3 + a = 1 - a \Rightarrow 2a = 4$

$$\Rightarrow a = 2$$

substituting in (3), $1 = c$ and using (1), $b = 3$ thus $a = 2$, $b = 3$, $c = 1$

Q4. AEB.

$$4(\sqrt{3} - i) = 8\left(\frac{\sqrt{3}}{2} - \frac{1}{2}i\right) = 8\left(\cos\left(-\frac{\pi}{6}\right) + i\sin\left(-\frac{\pi}{6}\right)\right)$$

$$= 8e^{-i\pi/6}$$

If $z^3 = 8e^{-i\pi/6} = 8e^{+i(2n - (1/6))\pi}$ for n an integer.

then $z = 2e^{\pi i(2n - (1/6))/3}$

so that $n = 1$, $z_1 = 2e^{11\pi i/18}$;

$n = 0$, $z_2 = 2e^{-\pi i/18}$;

$n = -1$, $z_3 = 2e^{-13\pi i/18}$ where $r > 0$, $-\pi < \theta \leq \pi$

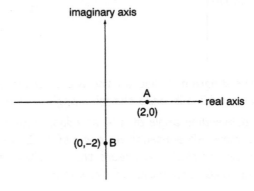

imaginary axis

A
(2,0)

real axis

(0,−2) B

Hence, equation arg $\left(\dfrac{z - 2}{z + 2i}\right) = \dfrac{\pi}{4}$,

can be rewritten $\arg(z - 2) - \arg(z + 2i) = \dfrac{\pi}{4}$ (1)

where, if $P(x,y)$ is any point on this locus, where $z = x + iy$, then $\arg(z - 2)$ means the angle between the line PA and the line AF $= \theta_1$ and $\arg(z + 2i)$ means the angle between the line PB and the line BG $= \theta_2$

Thus (1) becomes $\theta_1 - \theta_2 = \dfrac{\pi}{4}$ (see diagram below)

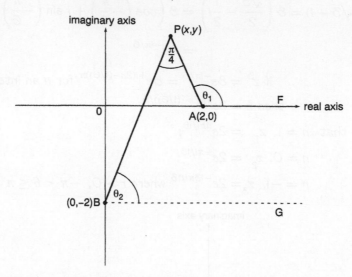

and from the diagram, P lies on the arc of a circle through AB, where the chord AB subtends the fixed angle $\pi/4$ at P.

This circle, subtending angle of $\pi/4$ at edge, must have its centre at origin O, where AB subtends an angle of $\pi/2$. Therefore P must lie on major arc of circle, to ensure that $\angle APB = \pi/4$. Centre of circle is origin $(0,0)$ and radius of circle is 2.

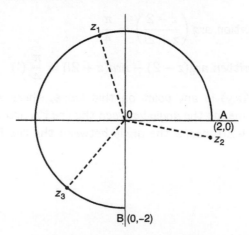

z_1, z_2 and z_3 all have moduli 2 and therefore all lie on the circle centre O, radius 2.

However, these points have arguments, $11\pi/18$, $-\pi/18$, $-13\pi/18$, and the point z_2, with argument $-\pi/18$, lies on the minor arc AB of the circle, and therefore is **not** on the above locus.

z_1 and z_3 are on the locus — see the above diagram.

Q5. AEB.

$$3u + v = 4i \tag{1}$$

$$u - iv = 8 + 2i \text{ thus } 3u - 3iv = 24 + 6i \tag{2}$$

(1) − (2) gives $v + 3iv = 4i - 24 - 6i$

$$v(1 + 3i) = -24 - 2i$$

$$\text{thus } v = -2\frac{(12 + i)}{(1 + 3i)} = \frac{-2(12 + i)(1 - 3i)}{(1 + 3i)(1 - 3i)}$$

$$\text{and } v = \frac{-2(12 + 3 + i - 36i)}{1 + 9} = \frac{-2(15 - 35i)}{10}$$

$$= -3 + 7i$$

so that $u = 8 + 2i + iv = 8 + 2i - 3i - 7 = 1 - i$

$u = 1 - i$ and $v = -3 + 7i$

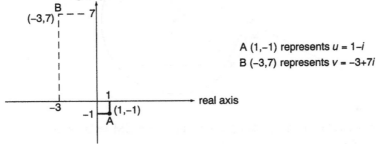

A (1,−1) represents $u = 1-i$
B (−3,7) represents $v = -3+7i$

P is the point (x,y), representing the complex number $z = x + iy$ when $|z - u| = |z - v|$, this means: **method 1**
the distance between (x,y) and $(1,-1) = $ distance between (x,y) and $(-3,7)$
i.e. PA = PB
and the locus of P is therefore the perpendicular bisector of the line AB.
i.e. line through mid point M $\left(\dfrac{-3 + 1}{2}, \dfrac{7 - 1}{2}\right)$,

$$\text{with gradient } \frac{-1}{\dfrac{7 - (-1)}{-3 - 1}} = \frac{-1}{8/-4} = \frac{1}{2}$$

Line is $y - 3 = \dfrac{1}{2}(x + 1)$

$$2y = x + 7$$

or, method 2

if you prefer: $|z - u| = |z = v|$ can be interpreted as follows
$$|x + iy - (1 - i)| = |x + iy - (-3 + 7i)|$$

i.e. $|(x-1)+i(y+1)| = |(x+3)+i(y-7)|$

$(x-1)^2+(y+1)^2 = (x+3)^2+(y-7)^2$ by definition

$\cancel{x^2}-2x+1+\cancel{y^2}+2y+1 = \cancel{x^2}+6x+9+\cancel{y^2}-14y+49$

$16y = 8x+56$

$2y = x+7$

Continuing method 1:

$|z-4-3i| = 5$ means the locus of P such that the distance between (x,y) and (4,3) is 5, i.e. a circle centre (4,3), radius 5, i.e. $(x-4)^2+(y-3)^2 = 25$

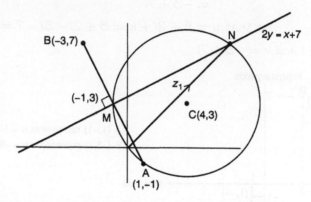

This circle must go through M because CM = 5, and M is point (−1,3).

The other point N is not so obvious, and so must be calculated, using equations of line $2y = x+7$ and circle $(x-4)^2+(y-3)^2 = 5^2$.

Solving these, $(2y-7-4)^2+(y-3)^2 = 25$

$5y^2-50y+130 = 25$

$\Rightarrow 5y^2-50y+105 = 0$

$\Rightarrow y^2-10y+21 = 0$

i.e. $(y-7)(y-3) = 0 \Rightarrow y = 7, 3 \Rightarrow x = 7, -1.$

Thus N(7,7) and M(−1,3)

$z_1 = 7+7i$ and $\arg z_1 = \dfrac{\pi^c}{4}$

$z_2 = -1+3i$ and $\arg z_2 = \arctan-\dfrac{3}{1} = 1.893^c$

(2nd quadrant — see diagram)

Printed and bound by CPI Group (UK) Ltd, Croydon, CR0 4YY

03/10/2024

01040338-0010